DO 0207547 4

UNIVERSITY
LIBRARY
WITHDRAWN

THE CODESIGN OF EMBEDDED SYSTEMS: A UNIFIED HARDWARE/SOFTWARE REPRESENTATION

THE CODESIGN OF EMBEDDED SYSTEMS: A UNIFIED HARDWARE/SOFTWARE REPRESENTATION

by

SANJAYA KUMAR

Honeywell Technology Center

JAMES H. AYLOR
BARRY W. JOHNSON
WM. A. WULF

University of Virginia

KLUWER ACADEMIC PUBLISHERS
Boston / Dordrecht / London

Distributors for North America:
Kluwer Academic Publishers
101 Philip Drive
Assinippi Park
Norwell, Massachusetts 02061 USA

Distributors for all other countries:
Kluwer Academic Publishers Group
Distribution Centre
Post Office Box 322
3300 AH Dordrecht, THE NETHERLANDS

Library of Congress Cataloging-in-Publication Data

A C.I.P. Catalogue record for this book is available
from the Library of Congress.

Printed on acid-free paper.

Printed in the United States of America

to our families,

whose love and support made this work possible

Table of Contents

List of Symbols and Acronyms

A	a set of hardware/software alternatives for a function
A_k	the kth hardware/software alternative of a function
B	a set of system blocks
C_j	a set of constraints for node j
C	a description of the communications paths between software and/or hardware units **or** the cost of a hardware/software alternative
Cm	a controlling mechanism
D_p	a decomposition of a function down to some level p
DG	a decomposition graph
e_{ij}	an edge from node i to node j within a graph
E	a set of edges within a graph
e_k	element k

f	a function
F	a set of functions
F^h	a set of hardware functions
F^s	a set of software functions
G	a set containing goodness values, grouped by metric
glb	greatest lower bound
H	a set of hardware units **or** the physical devices within a system
HM	a hardware model
HSM	a hardware/software model
I	a hardware/software implementation **or** an interpreter
I^h	a hardware interpreter
I^s	a software interpreter
l	a level
l^a	a level of abstraction
l^i	a level of interpretation
l^h	an encapsulation of both level of abstraction and level of interpretation for the hardware model: (l^a, l^i)
l^s	an encapsulation of both level of abstraction and level of interpretation for the software model: (l^a, l^i)
l^v	an encapsulation of both level of abstraction and level of interpretation for a virtual machine: (l^a, l^i)
L	a lattice **or** the set of leaf nodes within a decomposition
lub	least upper bound

M	a model
n_i	node i within a graph
N	a set of nodes within a graph
n^h	a hardware node within a graph
n^s	a software node within a graph
P	a program to be interpreted
Q	quantitative evaluation model
R	reliability
S_i	operator sensitivity of node i
S	a set of software units **or** a set of storage variables
SM	a software model
T	execution time
T_i	execution time of node i
U	a set of functions to be decomposed
v_i	virtual instruction i
V	virtual instruction set, a set of unique leaf functions associated with a decomposition
V_{HM}	set of operations supported by a hardware model
V_{SM}	set of virtual instructions associated with a software model
VM	virtual machine
W	a set of weights
Z	a set of functions which are not to be decomposed further
$\|X\|$	the cardinality of a set X

γ	a functional description used for hardware/software partitioning
γ_N	the nodes of γ
γ_E	the edges of γ
κ	the quality of a hardware/software alternative
λ	a set of goodness functions
$\rho(A)$	a set of metric functions
$\Pi(\gamma_N)$	a set partition of the nodes in γ
$\mathcal{P}_\gamma$	a hardware/software partition of the functional description γ
σ	an evaluation function which computes κ
ψ	a system description
ψ^F	a system description consisting of system functions
ψ^B	a system description consisting of system blocks
ψ^I	a system description consisting of hardware/software implementations
Γ	a hardware/software trade-off function
μ_i	the ith software unit
ν_i	the ith hardware unit

ADAS	Architecture Design and Assessment System
ADEPT	Advanced Design Environment Prototyping Tool
ADT	Abstract Data Type
ALU	Arithmetic Logic Unit
ASIC	Application Specific Integrated Circuit

CAD	Computer Aided Design
CPN	Colored Petri Net
CSIS	Center for Semicustom Integrated Systems (University of Virginia)
DA	Design Architect (Mentor Graphics Schematic Capture System)
FIFO	First In First Out
FSM	Finite State Machine
IPC	Interprocessor Communication
ISA	Instruction Set Architecture
PDL	Process Design Language
PMS	Processor Memory Switch
RISC	Reduced Instruction Set Computer
RMA	Rate Monotonic Analysis
ROM	Read Only Memory
SRS	Software Requirements Specification
TMR	Triple Modular Redundancy
VHDL	VHSIC (Very High Speed Integrated Circuit) Hardware Description Language
VHLL	Very High Level Language
VHSIC	Very High Speed Integrated Circuit
VLSI	Very Large Scale Integration

Preface

It is desirable to improve the development of complex systems, particularly embedded systems, so that products result which satisfy required constraints (cost, performance, reliability, and schedule, among others). Many believe that hardware/software codesign (or simply codesign), a more synergistic approach to system design, can aid in achieving this objective.

In the last five years, there has been increased interest in the area of codesign. The area of codesign has attracted an international community of individuals from academia and industry. These individuals represent several disciplines: systems engineering, system architecture, software engineering, and hardware engineering.

But, what exactly is codesign, and how is it different from the current hardware/software design process? Several perspectives exist as to what constitutes codesign. These perspectives range from supporting the existing hardware/software design process using existing technologies to providing fundamentally different design approaches. Areas of investigation include developing algorithms for performing hardware/software partitioning, modeling hardware/software systems, evaluating hardware/software trade-offs, and exploring the cross fertilization of techniques, such as the application of concepts from the hardware domain to the software domain and vice versa. This monograph presents one such perspective, bringing together several hardware and software ideas. The work was performed at the University of Virginia.

The monograph consists of ten chapters. The first five chapters are intended to be relatively broad, addressing the general area of codesign. The next four chapters focus on unified representations for hardware and software. In these chapters, coverage is given to the modeling of hardware/software systems, performance evaluation, the exploration of hardware/software trade-offs, and the notion of cross fertilization. The last chapter contains concluding remarks. Although some material is included which discusses the hardware/software partitioning problem, no new algorithms are presented. An overview of the individual chapters is provided below.

Chapter 1 begins by describing the motivation for codesign. The current design process and its consequences are also discussed. Some system design issues are mentioned as well. Next, the scope and goals are outlined. Finally, the organization of the remainder of the monograph is presented.

The goal of Chapter 2 is to furnish background for later chapters. The topics include embedded systems, models of design representation, the virtual machine hierarchy, performance modeling, and hardware/software development.

Chapter 3 contains an overview of codesign work, both past and present. Although most of the current work in codesign is new, many earlier works can also be considered codesign efforts. An informal view of codesign is first presented. This section is followed by a brief discussion of hardware/software trade-offs. The idea of cross fertilization between the hardware and software domains is then introduced. The next three sections focus on a typical codesign process, some codesign environments, and their limitations. The last section describes ADEPT (Advanced Design Environment Prototyping Tool), a VHDL-based environment for integrated performance and reliability evaluation being extended to support codesign. This environment is used to demonstrate several codesign ideas.

Chapter 4 designates the starting point of new codesign research within this monograph. To provide a common base for subsequent discussions, several important concepts are defined. The concepts include functions, functional decomposition, virtual machines and virtual instruction sets, hardware/software partitioning, hardware/software partitions, hardware/software alternatives, hardware/software trade-offs, and codesign. The codesign section discusses the exploration of hardware/software trade-offs and introduces a linear, weighted model for evaluating hardware/software alternatives with respect to multiple metrics. Finally, an example is presented to illustrate the weighted model.

In Chapter 5, a codesign methodology is described which supports the concepts developed in Chapter 4. This chapter starts by examining the amount of unification present in various hardware/software design approaches and stating the basic philosophies embodied by the codesign methodology. A framework for hardware/software codesign is then provided, which is used to guide important areas of investigation. The chapter closes with an example illustrating various aspects of the methodology.

Starting with Chapter 6, the focus of the monograph shifts to hardware/software modeling. The opening section discusses the idea of a unified representation for hardware and software. Several modeling concepts, such as level of abstraction and level of interpretation, are introduced. A unified representation, referred to as the decomposition graph, is then developed. The decomposition graph incorporates descriptions based on either functional abstractions or data abstractions. The final section concludes with related work.

In Chapter 7, an abstract hardware/software model is presented which utilizes a unified representation based on functional abstractions employing data/control flow concepts. The model supports early evaluation and attempts to address the current separation between the software and hardware design processes. First, the requirements and the applications of the model are described. Various models utilized for hardware/software systems are mentioned. The abstract hardware/software model is formalized, and its implementation is discussed. An example is used to illustrate the model, and the model's generality is explored. Related work is the subject of the last section.

Chapter 8 is an extension of Chapter 7. The applications of the abstract hardware/software model are revisited. Several examples are provided to demonstrate how the model can be used for early evaluation, namely general performance evaluation, identification of bottlenecks, hardware/software trade-off evaluation, and alternative evaluation. The examples include a railway control system, a system for aluminum defect detection and classification, a stylus tracking system, and a distributed system for supporting parallel discrete event simulation.

In Chapter 9, a unified representation based on data abstractions is used to model hardware components. Also, data decomposition, a decomposition technique based on abstract data types, is utilized for refinement. These ideas provide the basis for the application of object-oriented techniques to hardware design, which is an example of cross fertilization. The opening section motivates the use of object-oriented techniques. A brief discussion of data types is provided. Next, the modeling of hardware components using data abstraction and the use of inheritance for deriving specialized components are illustrated. The technique of data decomposition is presented and is demonstrated on a processor example. The notion of type genericity is briefly mentioned. Related work completes this chapter.

Chapter 10 summarizes the new codesign efforts described within this monograph (Chapter 4 through Chapter 9). The chapter outlines future work and then closes with some concluding remarks.

In addition to the references cited at the end of the monograph, the notes and handouts from the workshops listed below are good sources of information.

- 1991 Workshop on Hardware/Software Codesign, Austin, Texas, May 1991.
- CODES International Workshop on HW/SW Codesign, Grassau, Germany, May 1992.
- International Workshop on Hardware-Software Co-Design, Estes Park, Colorado, USA, September 1992.
- Codes/CASHE '93, 2nd IFIP International Workshop on Hardware/Software Codesign, Innsbruck, Austria, May 1993.
- International Workshop on Hardware-Software Co-Design, Cambridge, Mass., USA, October 1993.
- Proceedings of the 3rd International Workshop on Hardware/ Software Codesign, Grenoble, France, September 1994.

The following references are special issues.

- IEEE Computer, Hot Topics, January 1993.
- IEEE Design & Test, D & T Roundtable, March 1993.
- IEEE Design & Test, September 1993.
- IEEE Computer, Computing Practices, December 1993.
- IEEE Design & Test, December 1993.
- IEEE Computer, Computing Practices, January 1994.
- IEEE Micro, August 1994.
- IEEE Computer, February 1995.

We hope that this monograph provides some additional perspectives in the area of codesign. More importantly, it is our hope that this work generates further interest and encourages other individuals to explore this rapidly emerging area. Comments on the material are welcome. Our email addresses are provided below.

Sanjaya Kumar
skumar@src.honeywell.com

James H. Aylor
jha@virginia.edu

Barry W. Johnson
bwj@virginia.edu

Wm. A. Wulf
waw3s@virginia.edu

Acknowledgments

Several individuals have contributed to the efforts described in this monograph. These individuals include Ronald D. Williams, Ronald Waxman, Robert H. Klenke, Joanne B. Dugan, Stephen H. Jones, John C. Knight, Arne Bard, Ramesh Rao, Gnanasekaran Swaminathan, Maximo Salinas, Richard MacDonald, Moshe Meyassed, Eric Cutright, Sanjay Srinivasan, Charles Choi, Robert McGraw, Bill Miller, Anup Ghosh, Paul Perrone, Anees Shaikh, Sudhir Srinivasan, Carmen Pancerella, Peter Schaefer, Darrell Kienzle, Phil Smith, Lori Kaufman, and Shivani Kumar. Thanks to Mike Casey at Kluwer for his patience and understanding. We would also like to thank the Honeywell Technology Center for use of their computing facilities during the preparation of this monograph, particularly Fred Rose, Raj Kant, and the creative resources group.

This work has been supported by the Semiconductor Research Corporation and the National Science Foundation.

THE CODESIGN OF EMBEDDED SYSTEMS: A UNIFIED HARDWARE/SOFTWARE REPRESENTATION

Chapter 1

Introduction

Hardware/software codesign, the cooperative design of hardware and software, offers new investigations into several old problems in system design. In many circumstances, previous solutions to these problems have employed ad hoc techniques, producing unsatisfactory results. Some problems cannot be solved given the current approach to system design. However, the advent of new technologies and design environments presents opportunities for providing better, more structured solutions.

Hardware/software codesign brings together concepts and ideas from three primary disciplines within system design: system level modeling, hardware design, and software design. Research in this area addresses such aspects as the unification of the currently separate software and hardware development paths and the *exploration of hardware/software trade-offs*. Hardware/software trade-off exploration attempts to satisfy a set of objectives through the allocation of functionality into software and hardware.

This chapter starts by discussing the motivation for research in the area of hardware/software codesign. A brief presentation of some considerations in system design is then provided. Next, the scope and goals of the monograph are outlined. This discussion is followed by an overview of the remaining portions of the monograph.

1.1 MOTIVATION FOR HARDWARE/SOFTWARE CODESIGN

Several myths [1] regarding the development of the hardware and software of complex computer systems, particularly embedded systems, persist today. It is believed that hardware and software can be developed independently and successfully integrated later. Because of software's "malleable" nature, it is also believed that hardware inadequacies can be easily rectified by simple software changes. Another commonly held belief is that software, once accepted operationally, never requires change. Lastly, it is a simple task to state valid and complete software requirements and then implement them.

Referring to Figure 1.1, these myths are embodied in the existing design process for hardware/software systems [2][3]. Current practice dictates the separation of the hardware and software development paths early in the design cycle. These paths remain independent, with very little interaction occurring between them until system integration. In many circumstances, a "hardware first" approach is adopted [1][4]. In such an approach, hardware is specified without fully appreciating the computational requirements of the software in terms of processor speed and memory capacity. In addition, software development does not influence hardware development and does not track changes made during the hardware design phase [2]. It is finally during system integration that the software and hardware are combined and tested as a whole.

The use of such a methodology has several consequences [2][5]. Because integration occurs late in the process, problems encountered at this time may require modification of the software and/or hardware, resulting in potentially significant cost increases and schedule overruns. For example, the premature selection of hardware may require that the software attempt to correct hardware inadequacies [6]. As shown in Figure 1.2, a phenomenon that has been observed [1][7] as a result of this "hardware first" approach is the tripling of relative programming cost per instruction when working with a 90% saturated machine (in terms of processing speed and memory capacity). Also, poor software performance may necessitate the development of additional hardware late in the process [8].

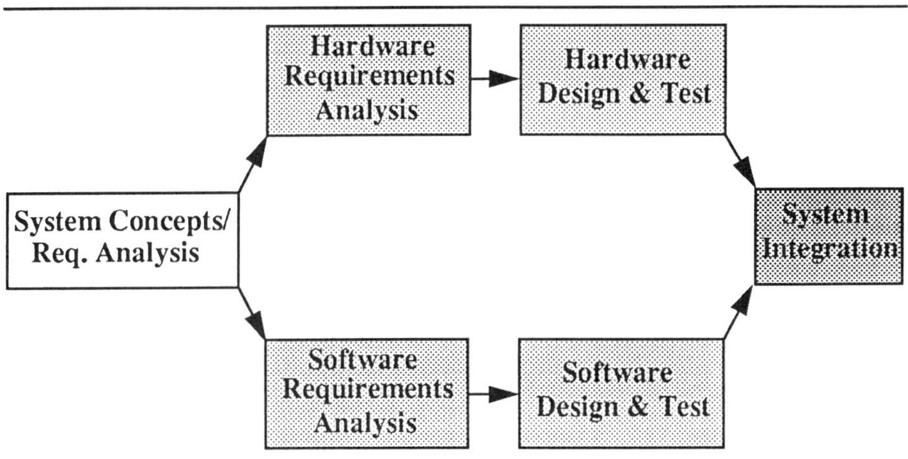

Figure 1.1 Current system development methodology

Problems encountered at a late stage of software development affect both cost as well as schedule. Specifically, problems detected during development testing are at least an order of magnitude more costly to fix than those detected during requirements [9][10]. Some surveys [11] of software projects have indicated an average cost overrun of 33%-36% and an average schedule overrun of 22%. Almost 50% of the respondents in one of the surveys attributed the cost overrun to frequent changes in design and implementation. Late hardware modifications exacerbate these problems.

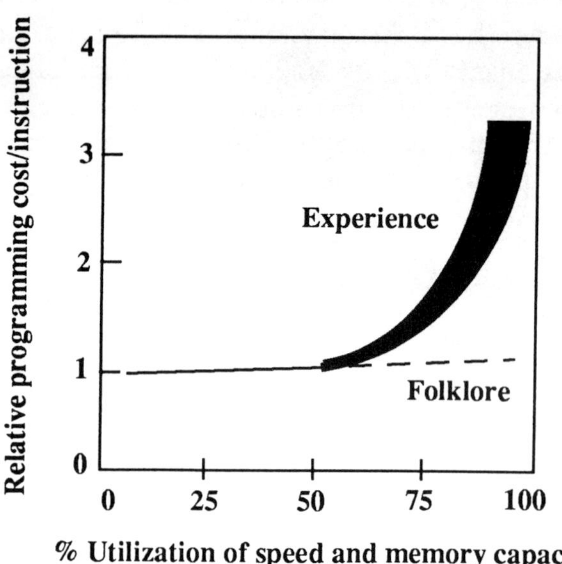

Figure 1.2 Impact of inadequate hardware resources (From [1], © 1978 IEEE)

Another consequence is that the ability to explore hardware/ software trade-offs is restricted, such as the movement of functionality from the software domain to the hardware domain (and vice-versa) or the modification of the hardware/software interface. This restriction does not allow hardware/software inadequacies to be corrected as the development proceeds. Thus, the inability to explore hardware/ software trade-offs affects the quality of the resulting system implementation.

The advent of certain technologies has changed design methodologies for computer systems. For example, improvements in electronic design automation tools, such as high-level hardware synthesis capabilities [12], and application specific integrated circuit (ASIC) development allow complex algorithms to be implemented in silicon quickly and inexpensively. On the other hand, the emergence of reduced instruction set computer (RISC) technology [13] has allowed functionality within the processor to be transferred into software. The result of such technologies is the "blurring" of decisions regarding the implementation of functionality in hardware versus software. Therefore, the decision to allocate functionality in hardware versus software is not as straightforward, and applications require a case-by-case analysis [3].

Because of the problems and new technologies described above, a more unified, cooperative approach to the design of hardware/software systems is required, one in which the hardware and software options can be considered together [3]. This approach is called *hardware/software codesign*, or simply *codesign* [3][14][15]. Codesign leads to more efficient implementations and improves overall system performance, reliability, and cost effectiveness [15]. Also, because decisions regarding the implementation of functionality in software can impact hardware design (and vice-versa), problems can be detected and changes made earlier in the development process [16]. Because software costs dominate hardware costs [9], some have suggested that software requirements should drive the hardware design of a system [16].

Codesign can especially benefit the design of *embedded systems*, systems which contain hardware and software tailored for a particular application. As the complexity of these systems increases, the issue of providing design approaches that scale up to more complicated systems becomes of greater concern. A detailed description of a system can approach the complexity of the system itself [17], and the amount of detail present can make analysis intractable. Therefore, decomposition techniques and abstractions are necessary to manage this complexity.

1.2 System Design Considerations

The process of system design entails several activities: system level modeling, hardware design, and software design. System level modeling, which addresses concerns such as performance, occurs in isolation of other engineering disciplines [18]. This isolation has led to the *model continuity problem* [3], the inability to gradually refine a system level model into a hardware/software implementation. In most circumstances, the system level model is not used in later stages of the design process. Thus, the association between the system level model and the hardware/software implementation is lost.

Providing model continuity can help address many fundamental problems in system design. For example, it has been recognized that many complex systems do not perform as expected in their operational environment [1][17]. Model continuity would allow the validation of system level models with their corresponding hardware/software implementations. Model continuity is also important because hardware/ software trade-offs can be performed at several stages of the design process.

Another problem in system design is the failure to appreciate the subtleties associated with integrating subsystems [1][19][20]. The well known 90/50 rule states that although 90 percent of the time ASICs work the first time, only 50 percent work properly in the system. This problem applies to the integration of hardware/software subsystems as well. Finally, model continuity can help test the assumption that optimization of individual subsystems will automatically lead to optimization of the entire system [17].

These problems have prompted the need for a unified design environment [21][22], free of multiple languages, translators, and simulators, that supports system level modeling, hardware/software codesign, and model continuity. The construction of a unified design environment requires a common representation for performing these design activities [23][24][25]. A common representation helps to address the problem of model continuity by allowing system level models to gradually evolve into hardware/software implementations. In addition, the use of such a representation eases the task of performing hardware/software trade-offs as the system is being designed.

1.3 Research Scope and Overview

There are many problems to be solved in hardware/software codesign of which only a subset is addressed in this monograph. One of the most heavily researched problems is *hardware/software partitioning*, deciding which functions should be implemented in hardware and which in software. Although the monograph presents some thoughts on this topic, no new algorithms for performing hardware/software partitioning are provided. Requirements analysis and test are important activities in the design process. However, these design phases are also given only minor treatment.

Figure 1.3 and Figure 1.4 summarize the primary areas of focus in this monograph. One goal is to provide a unified view of hardware and software. This goal is addressed through the development of a *unified representation*, a representation that can be used to describe hardware or software. A premise of this representation is that hardware and software are similar, and thus, there is no fundamental difference between the two.

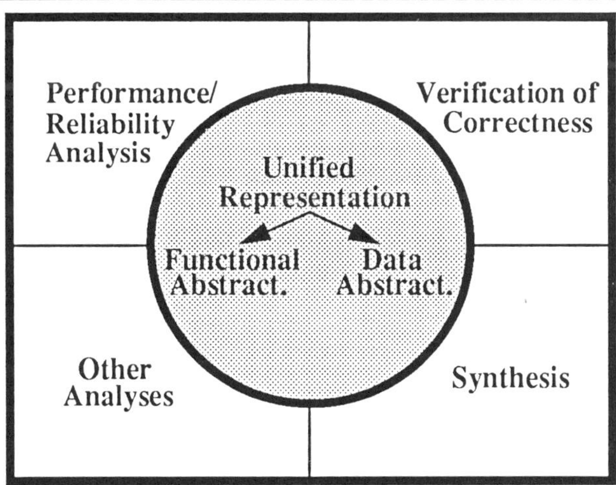

Figure 1.3 Benefits of a unified representation

The similarity between hardware and software is a common theme that appears throughout the monograph in several forms. In addition to providing a common paradigm for hardware/software development, the benefits of a unified representation include the ability to support several design techniques and analyses in a uniform fashion for hardware and software. The unified representation developed in this monograph incorporates descriptions based on either functional abstractions or data abstractions. As an example of functional abstractions, data/control flow representations were explored for performance analysis. Descriptions based on data abstractions led to investigations in the application of object-oriented techniques to hardware design.

As indicated in Figure 1.4, an abstract hardware/software model has been developed to promote early performance analysis. Using a unified representation based on data/control flow concepts, the abstract hardware/software model supports general performance evaluation, the identification of bottlenecks, the evaluation of hardware/software trade-offs, and the evaluation of design alternatives. This model can be utilized to assess the consequences of various hardware/software partitioning decisions before committing to a particular design.

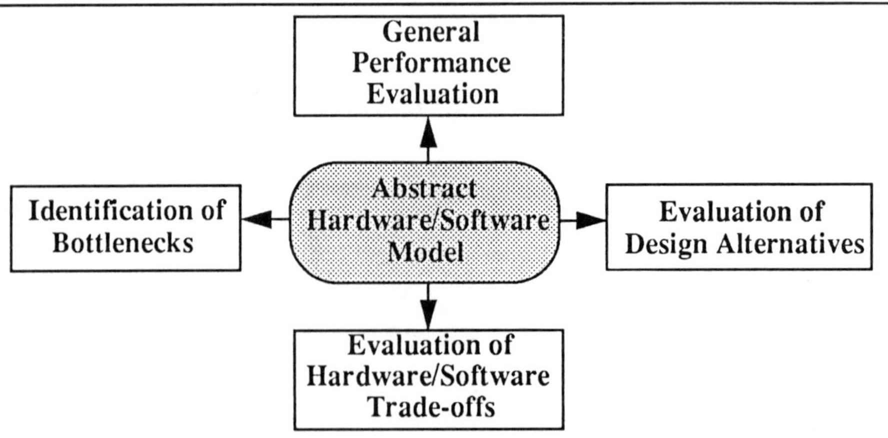

Figure 1.4 Applications of an abstract hardware/software model

1.4 A ROAD MAP OF THE MONOGRAPH

A road map for the remaining portions of the monograph is illustrated in Figure 1.5. Chapter 2 contains background material. The intent of this chapter is to prepare the reader for subsequent discussions on hardware/software codesign. After obtaining an understanding of this material, Chapter 3 provides an overview of hardware/software codesign. This chapter includes a discussion of topics that are relevant to codesign and surveys both past as well as present efforts in this area. A unified modeling environment currently under development is also briefly described. This environment is used as a vehicle for subsequent codesign explorations.

The remaining chapters of the monograph reflect new codesign investigations. In Chapter 4, some important codesign concepts are presented. In Chapter 5, a new codesign methodology which supports the concepts in Chapter 4 is described. Although a complete implementation of the methodology is not provided, this discussion provides further motivation for the abstract hardware/software model.

A unified representation, incorporating both functional abstractions and data abstractions, is the subject of Chapter 6. Before presenting the unified representation, this chapter develops concepts that are relevant to hardware/software modeling. A description of an abstract hardware/software model for early performance evaluation is the focus of Chapter 7. This model employs a unified representation based on functional abstractions. The application of the abstract hardware/software model is demonstrated using a variety of examples in Chapter 8. Techniques for identifying bottlenecks and performing hardware/software trade-offs are also presented in this chapter. Using a unified representation based on data abstractions, Chapter 9 discusses the application of object-oriented techniques to hardware design. Finally, a summary of the new codesign explorations, future work, and concluding remarks are provided in Chapter 10.

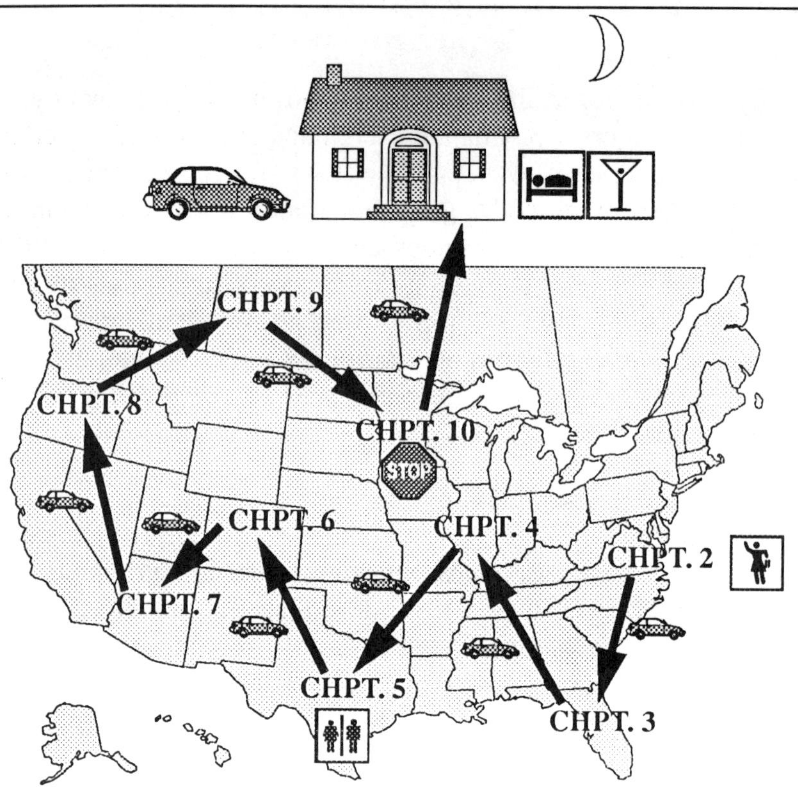

CHPT. 2 Hardware/Software Background
CHPT. 3 Hardware/Software Codesign Research
CHPT. 4 Codesign Concepts
CHPT. 5 A Methodology for Codesign
CHPT. 6 A Unified Representation for Hardware and Software
CHPT. 7 An Abstract Hardware/Software Model
CHPT. 8 Performance Evaluation
CHPT. 9 Object-Oriented Techniques in Hardware Design
CHPT. 10 Concluding Remarks and Future Work

Figure 1.5 Road map of the monograph

Chapter 2

Hardware/Software Background

To better appreciate hardware/software codesign research, this chapter introduces embedded systems, discusses models of design representation, describes the concept of a virtual machine hierarchy, examines performance modeling, and presents important aspects of hardware/software development. The intent of this chapter is not to provide a tutorial-like presentation of these topics but to highlight those areas that are relevant to hardware/software codesign.

2.1 EMBEDDED SYSTEMS

Codesign can benefit the development of *embedded systems* [26][27]. Embedded systems are application specific systems which contain hardware and software tailored for a particular task and are generally part of a larger system. General purpose processors, special purpose processors, and ASICs are among the many components used to implement these systems. Embedded systems often fall under the category of reactive systems [28], systems which continuously interact with humans and their environment, with real-time systems being reactive systems that must meet some time constraint [26][29][30].

Embedded systems abound in everyday life [31]. Examples include the engine control unit of an automobile and an aircraft autopilot. These systems are also found in process monitoring and control, signal processing, home appliances, industrial robots, and laser printers.

Typical metrics that impact the design of embedded systems include reliability, performance, cost, and form factor, which includes size, weight, and power constraints [31][32]. Reliability is an issue in systems which have to perform unattended for long periods of time, such as a satellite, or in which the consequences of a failure can be disastrous, such as an aircraft autopilot. In critical applications, different implementations of a software function may be developed to run on different processors.

Embedded systems can be divided into two broad classes based on performance [32][33]. Low to moderate performance systems have severe cost and form factor requirements. Examples include controllers for home appliances. For these applications, microcontrollers are typically sufficient. High performance systems require more powerful microprocessors, and in some cases, multiprocessor systems are employed. Fly-by-wire systems [34] in avionics and robots with tactile and vision systems are examples of such applications. In some systems, independent software tasks may be required to execute on demand, requiring the use of an operating system.

To appreciate the complexity of these systems, Figure 2.1 shows the flight control system for the X-29 aircraft [34]. The flight control system consists of three computers arranged in a triple modular redundancy (TMR) configuration. Inputs are received from sensors and cockpit controls/displays. Using these inputs, the computers perform identical computations. Once the computations are performed, a majority vote of the outputs is formed, and commands are then issued to the control surfaces (ailerons and rudders).

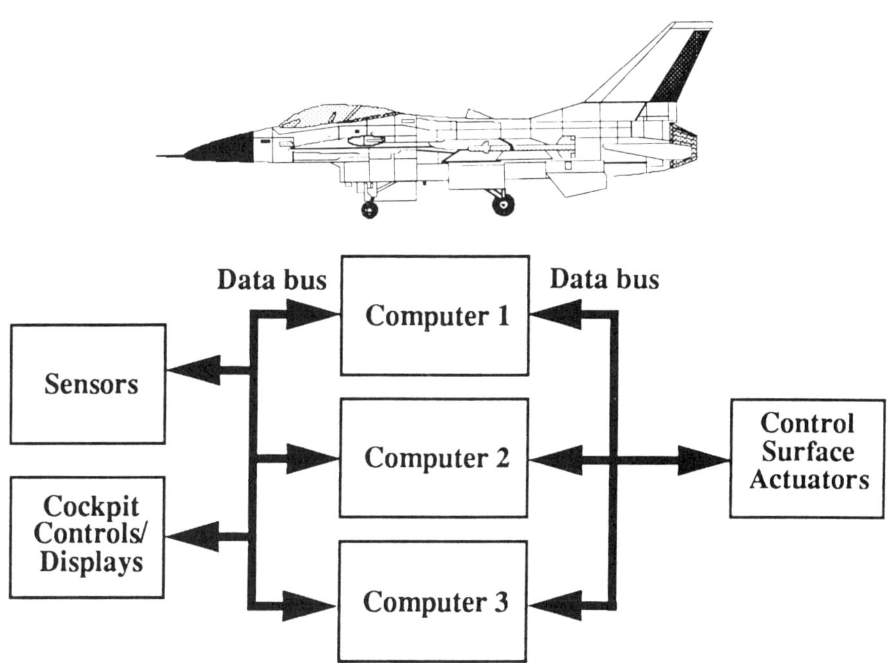

Figure 2.1 Flight control system for X-29 aircraft (From [34], adapted from pg. 12, © 1989 by Addison-Wesley Publishing Company, Inc. Reprinted by permission of the publisher.)

2.2 MODELS OF DESIGN REPRESENTATION

Models of design representation describe design activities at various levels of detail [12]. Many of the models that exist today are based on the design hierarchy described by Bell and Newell [35][36]. Since then, several models for digital systems have been presented in the literature. These models include the Gajski-Kuhn Y-Chart [37] and the Walker-Thomas model [38]. Many aspects of these models are similar, and the reader is referred to the above references to better understand their differences.

A model based on the representation described by McFarland et al. [12], which shares much in common with the Walker-Thomas model [38], is presented in Table 2.1. In this model, digital systems are described at the system, algorithmic (instruction set), register-transfer, logic, and circuit levels of abstraction. In addition, the model describes three different domains, or views, of a system. These domains are behavioral, structural, and physical. A design process from concept to implementation is captured in this model by starting at the system level in the behavioral domain and finishing at the circuit level in the physical domain.

Table 2.1 Model of design representation (From [12], © 1990 IEEE)

Abstraction	Behavioral	Structural	Physical
System (PMS)	Information flow	Processors, Memories, Switches	Cabinets, Cables
Instruction Set (Algorithmic)	Input/output transforms	Processors, Memories, Ports	Board Floorplan
Register-Transfer	Register transfers	Registers, ALUs, Busses, Muxes	Integrated Circuits, Macro Cells
Logic	Boolean equations	Logic gates, Flip-flops	Standard Cells
Circuit	Network equations	Transistors	Transistor layout

At the system level, designers are concerned with information flow, such as data and control flow, and the design activities include performance and reliability modeling. Petri nets [39] and queuing models [40] are employed for performance analysis at this level. The hardware components consist of processors, memories, and switches

(PMS) [35] that manipulate packets, or information units, which flow through these components. Note that PMS is a notation for describing computer systems as well. Markov models [34] and fault trees [41] are commonly used for reliability analysis.

Hardware description languages, such as ISP [36], ISPS [42], and VHDL [43][44][45], or programming languages, such as "C" [46], are used to represent hardware at the algorithmic, register-transfer, and logic levels of abstraction. At the algorithmic level, processors and memories are described as input-output transformations which communicate through ports. In languages that support concurrency, the components are described as concurrent processes.

The primary hardware components at the register-transfer level are registers, arithmetic logic units, busses, and multiplexers. A sequence of register transfers initiated by a control unit allows data to be transferred between these components. At the logic level, gates and flip-flops are utilized. The behavior of these components is expressed using Boolean expressions. Transistors, whose behavior can be described through a collection of network equations, are employed at the circuit level.

2.3 The Virtual Machine Hierarchy

A deficiency of the model described in Table 2.1, which was addressed by Bell and Newell [35][36] to some extent, is that the many levels within software are not adequately represented. One reason for this is that the levels are not the same for all software systems. A typical computer system can be described in terms of the layers, or levels, illustrated in Figure 2.2 [47][48]. As will become apparent shortly, these layers describe a *virtual machine hierarchy*. The primary motivation for structuring a computer system in this manner is to manage complexity. Much of the discussion below can be found in [48].

Application developers utilize a computer system to help them solve problems. These users formulate a solution to a problem by means of a *program*, a sequence of instructions that describe how the problem is to be solved. The machine employed to execute a program understands a limited set of instructions, for example, the instructions specified by the *instruction set architecture* (*ISA*), the programmer's view of the machine. Thus, one way of solving a problem is to construct a program in terms of the instructions of the ISA and then to execute this program on the machine.

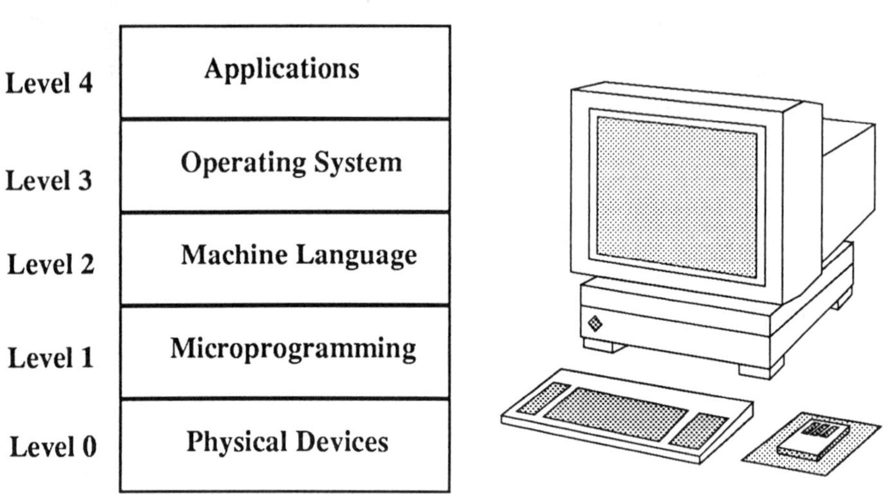

Figure 2.2 Layers within a typical computer system (From [47], © 1987, p. 2. Reprinted by permission of Prentice Hall, Upper Saddle River, New Jersey.)

However, for complex problems, it is tedious to develop a program in terms of only those instructions provided in the ISA. Therefore, it is desirable to provide a more convenient set of instructions to program with. This new set of instructions presents the application developer with a *virtual machine*, one that is easier to program than the ISA level machine. A virtual machine provides an abstraction, and a developer can treat the virtual machine as if it were the actual machine. For example, when using a high level language, such as "C", the developer writes programs for a virtual machine that understands the instructions of the language. One can continue creating more convenient "virtual instruction sets" and corresponding virtual machines.

As indicated in Figure 2.2, the lowest level consists of the physical devices within the computer system. Although not explicitly shown in the figure, this level can also be divided into several levels, such as the register-transfer level, the logic level, and the circuit level. The microprogramming level consists of primitive "software", usually located in a read-only-memory (ROM), that directly controls the physical devices. This software is also called firmware, and some individuals consider this level part of the hardware. In many computer systems, this level does not exist, and the control functions are provided by hardware.

The machine language level consists of programs written in terms of assembly language. This level is equivalent to the ISA view of a machine. The operating system level provides a higher level set of services. For example, the operating system may provide file abstractions which can be opened and closed, hiding the complexity of the detailed operation of a disk. Such abstractions provide the user with yet another more convenient virtual machine interface. Although not shown explicitly, in a manner similar to the hardware level, the operating system level can also be further divided into distinct levels [49]. The topmost level consists of the applications to be run on the computer system.

Therefore, each level in Figure 2.2 (with the exception of the level corresponding to the "real" machine) represents a virtual machine abstraction [50][51][52][53]. A virtual machine provides a set of facilities, such as operations and resources, and hides certain design decisions. An architecture can be defined for each level, containing a set of data types, operations, and other features visible to a user at that level. As an illustration, a machine language level programmer is cognizant of the available instructions, various data types supported, and memory structures. However, certain design decisions are hidden from the programmer, such as whether a microcoded or a hardwired control unit is used.

2.4 PERFORMANCE MODELING

Within the realm of system level modeling, performance issues are emphasized in this monograph. *Performance modeling*, also called uninterpreted modeling [21], is utilized in the early stages of the design process to analyze systems in terms of such metrics as throughput and utilization. Performance models are also used to identify bottlenecks within a system. The term "uninterpreted modeling" reflects the view that performance models generally lack valued data and functional (input-output) transformations. However, in some circumstances, this information is necessary to allow adequate analysis to be performed.

A variety of techniques have been employed for performance modeling. The most common techniques are Petri nets [39][54][55][56][57] and queuing models [40][58][59][60]. A combination of these techniques, such as a mixture of Petri net and queuing models [57][61], has been utilized to provide more powerful modeling capabilities. All of these models have mathematical foundations. However, models of complex systems constructed using these approaches can quickly become unwieldy and difficult to analyze [62].

Examples of a Petri net and a queuing model are shown in Figure 2.3. A Petri net consists of places, transitions, arcs, and a marking. The places are equivalent to conditions and hold tokens, which represent information. Thus, the presence of a token in the place of a Petri net corresponds to a particular condition being true. Transitions are associated with events, and the "firing" of a transition indicates that some event has occurred. A marking consists of a particular placement of tokens within the places of a Petri net and represents the state of the net. When a transition fires, tokens are removed from the input places and are added to the output places, changing the marking (the state) of the net and allowing the dynamic behavior of a Petri net to be modeled.

Petri nets can be used for performance analysis by associating a time with the transitions. Timed and stochastic Petri nets contain deterministic and probabilistic delays, respectively. These Petri nets are uninterpreted, since no *interpretation* (semantics) is associated with the transition or the tokens. This idea of associating semantics with various elements of a model will be explained in subsequent chapters.

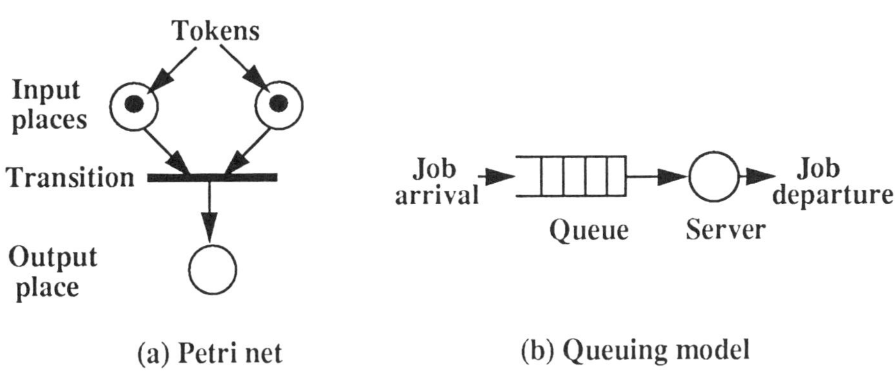

Figure 2.3 Petri net and queuing model

A queuing model consists of queues and servers. Jobs (or customers) arrive at a specific arrival rate and are placed in a queue for service. These jobs are removed from the queue to be processed by a server at a particular service rate. Typically, the arrival and service rates are expressed using probability distributions. There is a queuing discipline, such as first-come-first-serve, which determines the order in which jobs are to be serviced. Once they are serviced, the jobs depart and arrive at another queue or simply leave the system. The number of jobs in the queues represents the model's state.

Numerous environments exist which are based on Petri nets. One of the earliest was ADAS [15], which used timed Petri nets for performance analysis [63]. The uninterpreted/interpreted (performance/ functional) modeling methodology [21] is a graph-based approach which utilizes a collection of modules [64], each having a corresponding

VHDL and Petri net representation, to support performance modeling. Timed Petri nets, extracted from Occam constructs for performance analysis, are employed in a software tool called the System Design Workbench [65].

Several environments based on queuing models have also been developed. Simulation languages such as GPSS [66] and Simscript [67] have been used for performance modeling. However, the models constructed using these languages take considerable effort to write and debug. Also, models result which are difficult to maintain and modify [68]. Therefore, several graphical, queuing-based environments have been used in industry. These systems include IBM's RESQME [69], AT&T's PAW [70], and Q+ [71], an extension of PAW. However, these environments do not support modeling at lower levels of the design process. Some attempts to address both performance modeling and modeling at lower levels include N.2 [72] and the TD Technologies toolset [73].

Other graph-based approaches such as UCLA graphs [74] used in SARA [75] and SES/Workbench [76], which is based on Information Research Associate's PAWS [68] (not to be confused with AT&T's PAW above), also support queuing behavior. The UCLA graph model has a strong mathematical foundation and supports extensive analysis. SES/Workbench uses directed graphs constructed from a collection of high level primitives to model a system.

Directed graph representations have been used to analyze the performance [77][78][79][80][81][82][83][84] and reliability [85] of software. Queuing-based approaches, employing simulation and analytic techniques, have also been utilized for software performance modeling [86]. Techniques which combine several of these modeling approaches have been reported as well [87]. In addition, several efforts have focused on the development of software reliability models [88][89].

The notion of uninterpreted software [90][91][92] is analogous to the uninterpreted modeling found in hardware. These uninterpreted descriptions model the control structure of a program and can be converted to interpreted descriptions by specifying the functions in the software and associating meanings to the tokens. Petri nets, structured programming constructs, and data flow graphs have been used to represent uninterpreted software descriptions. Incorporation of temporal information, as in the case of timed Petri nets, allows these descriptions to be used for performance analysis.

The design of distributed systems requires considerable software analysis. Two fundamental problems in distributed data processing are task partitioning and task allocation [93]. These two problems are equivalent to software partitioning, breaking down an individual task into modules, and scheduling the modules onto processors, respectively. The goal of these steps is to maximize the performance of a system. Thus, issues such as interprocessor communication (IPC), the cost of communication between modules executing on separate processors, must be taken into account. Data partitioning, the partitioning of data across several processors, becomes important as well in order to improve performance by allowing concurrent operations to be performed on the data.

Task partitioning is considered a software design issue [93]. Approaches have been developed that aid in determining the number and size of the modules [94]. Environments have also been created to analyze the performance of various task partitions earlier in the design process [4]. Also, efforts have focused on the task allocation problem. Graph theoretic approaches have been utilized to determine the "optimal" mapping of modules onto processors [95][96][97]. In these approaches, software modules represent nodes in the graph, and the interprocessor cost for communication is placed on an edge connecting two nodes. The goal of most of these approaches is to minimize total cost in terms of execution time.

2.5 HARDWARE/SOFTWARE DEVELOPMENT

This section provides an overview of system development, software development, and hardware development activities.

2.5.1 System Development

Several descriptions of development methodologies and processes exist [2][3][6][34]. A commonly referenced description is the Department of Defense (DoD) Standard 2167 for system development [3]. The descriptions in this and subsequent sections are intended to bc relatively generic. However, at a high level, they follow the 2167 model outlined in [3]. It should be emphasized that the description which follows is not universal. The objective is to describe some of the major steps that are typical in the development of complex systems.

The simplified development methodology from Chapter 1 is shown in Figure 2.4 for convenience. One of the first stages of system development is problem definition [6][34]. The purpose of this stage is to develop functional and nonfunctional requirements for the system. Some system requirements may be expressed quantitatively in terms of performance, reliability, cost, and form factor. Other requirements may be stated qualitatively, for example, addressing compatibility or interfacing issues.

The next stage involves evaluating candidate designs and selecting the mixture of hardware and software to be used through hardware/ software partitioning. The result of this stage is a collection of hardware and software requirements. Performance modeling can be used to aid in the evaluation of alternative designs. Note that implementation technologies and economics are also considered. It should be mentioned that some iteration may be involved between this stage and the problem definition stage. With respect to Figure 2.4, the system concepts/ requirements analysis block encompasses both of these stages.

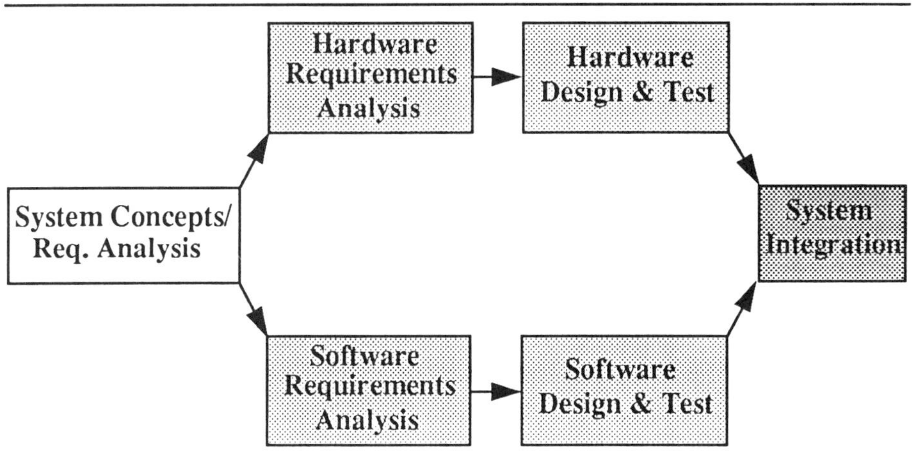

Figure 2.4 Current system development methodology

2.5.2 Software Development

This section is broken down into three primary parts. The first part presents an overview of the software development process. The second part focuses on abstractions used for software development. In the last part, a brief description of software synthesis techniques is provided.

Overview. The software life cycle consists of several phases [98][99]. Referring to Figure 2.5, the major phases of development are requirements, design, coding, and testing. Once testing is complete and the software has been accepted, the software reaches the operational phase. Although shown as steps with well defined boundaries, the point at which one phase ends and another phase begins is not as precise as suggested. Also, as will be explained shortly, these steps do not always occur in one direction.

The requirements phase consists of analysis and specification [99]. The result of this process is a software requirements specification (SRS) document which contains information regarding the functionality, performance, design constraints, and external interfaces of the software. Most SRSs are written in natural language. However, these descriptions result in documents that are ambiguous, incomplete, and inconsistent. A problem that often arises is the changing of requirements as the software is being developed, for example, due to new requirements imposed by the client.

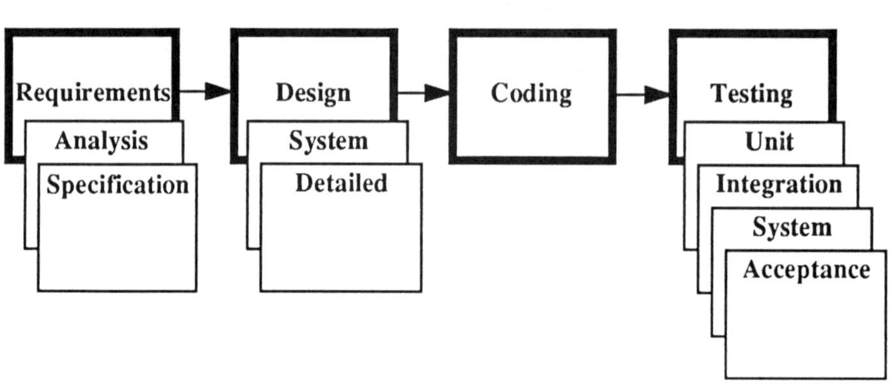

Figure 2.5 Major phases of software development

Because errors in SRSs are extremely costly to fix later in the design process [10], alternative techniques have been developed to specify a system's external behavior [99]. These specification techniques, many of which are executable [29], include state machines [100][101][102], decision tables, data flow diagrams [103][104], Petri nets [105][106], languages, such as PAISLey [107][108], and Statecharts [28][29][109]. Data flow diagrams are popular and can be expressed hierarchically.

The design phase consists of two primary components: system (top-level or preliminary) design and detailed design. Structure charts [110] are useful during the system design phase. These constructs help to define the functional modules within the software system and their interfaces (inputs and outputs). Module specifications, which treat the module as a black box, are also important during this phase. One formal method for specifying modules consists of using pre- and post-conditions [111]. Pre-conditions are logical assertions on the input state, and post-conditions are logical assertions on the output state. Thus, pre- and post-conditions reflect what is assumed to be true upon entry and exit from a module, respectively.

An example of a functional specification for a SORT module is shown in Figure 2.6 [98]. The pre-condition indicates that the list of integers to be sorted is expected to be nonempty. The post-condition states that all of the elements are to be sorted in ascending order. Several implementations, such as bubble sort or insertion sort, can satisfy this specification. As an aside, note that this specification is flawed. Nothing is stated about the requirement that the final state of the list should contain the same elements as the initial list. An improved specification would indicate that the elements of the final list are a permutation of the elements in the initial list.

```
SORT(L: list of integers)
pre-condition: non null L
post-condition: forall i, 1 <= size(L), L[i] <= L[i+1]
```

Figure 2.6 Functional specification (From [98], © 1991, p. 195. Reprinted by permission of Springer-Verlag.)

Historically, software development has emphasized program correctness. Pre- and post-conditions were initially developed to verify that an implementation satisfies the functional specification. For example, the axiomatic method proposed by Hoare [111] utilizes a collection of rules and axioms to prove various theorems about a program. Other techniques for verification of programs include code inspections, static analysis, and symbolic execution [98].

The detailed design phase involves developing the internal logic (implementation) for the modules defined in the system design phase. In many cases, the process design language (PDL) [112] is used for both of these phases, particularly the detailed design phase. Coding is usually performed in some high level language, such as "C" or "C++".

Testing consists of several stages as well [98][99]. These stages include unit testing, integration testing, system testing, and acceptance testing. Unit testing involves testing individual modules while integration testing focuses on several, interconnected modules. System testing examines the entire software subsystem in the intended hardware environment. Regression testing may be required if changes are necessary. Finally, acceptance testing is performed by the customer or the intended user of the software.

Several design process models exist which attempt to guide the development of software. These models suggest an overall process for software development and include the waterfall model [113], prototyping [98], iterative enhancement [114], and the spiral model [115]. The waterfall model assumes that the development phases in Figure 2.5 follow a sequential order. For some systems, this assumption is not valid. Prototyping is used to better understand a client's requirements by developing a temporary, rudimentary version of the system. Once the requirements are understood by both clients and developers, the prototype system is typically thrown away. In iterative enhancement, software is developed incrementally until the entire system is complete, allowing testing to be performed in a stepwise fashion. In the spiral model, design activities are arranged as a collection of cycles. Within every cycle, several activities are performed: determining the objectives of the current cycle, evaluating alternative approaches for reaching the desired objectives, risk analysis [115][116], developing strategies to address the risks, software development, and planning for the next stage. Risk analysis plays an integral role in this process model. As an example, detailed prototypes may be developed to evaluate performance concerns.

Abstractions. The two primary abstractions used in software development are *functional (procedural) abstractions* and *data abstractions* [98][117]. Functional abstractions provide the basis for structured design techniques. Specifications for functional abstractions describe input-output behavior, and thus, hide the complexity associated with an implementation of the function. In this context, implementation refers to a realization of the function in terms of algorithms and data structures.

For example, “square_root(num)” (see Figure 2.7) is a functional abstraction since it is treated as a “black box” with well defined inputs and outputs. This information is all that is required in order to use the function in a program with confidence. These abstractions have been utilized in structured programming [112][118][119][120][121], structured design [103][104][122][123][124][125], stepwise refinement [126], and top-down design [127][128].

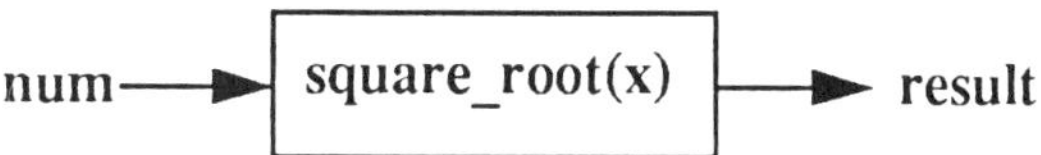

square_root(num: real) returns result: real
pre-condition: num >= 0
post-condition: num := result * result

Figure 2.7 Functional abstraction

A *data type* [129] consists of a domain of values and a set of operations. An example of a data type is the programming language type “integer”. The type integer has the domain of values $\{\ldots, -2, -1, 0, 1, 2, \ldots\}$ and the set of operations (among others) multiply, add, subtract, and divide. The latter defines the operations that can be performed on objects of type integer. Thus, two integers can be multiplied, added, and so on.

Data abstractions, or abstract data types [129][130], provide the basis for object-oriented design techniques. An *operational specification* (or *abstract model*) is used to define data types. Such a specification has two parts. The first part is a domain specification, consisting of an abstract (mathematical) description of the type. The second part is a description of the abstract operations (also called methods) that can be performed on objects of that type. An implementation (realization) for a data abstraction consists of a concrete representation for the data type and implementations for the operations. A concrete representation, such as a linked list, reflects a design decision on the part of the developer.

Parnas [131][132] pointed out the need to "hide" implementation information from the users of an abstraction, a concept called *information hiding*. Information hiding attempts to hide design decisions that are likely to change. Modules of large, complex designs are likely to change many times. In order to preserve the ability to change the implementation of a module without affecting its users, it is crucial that those users depend only upon the module's external specification and that all implementations preserve that specification.

Data abstractions were invented to provide the means for enforcing at least some of Parnas' s information hiding. Consider, for example, the abstract integer stack depicted in Figure 2.8. The stack includes the operations push and pop, and the shaded region depicts some (hidden) concrete representation.

The operational specification for the stack is shown in Figure 2.8. In the notation, the prime symbol indicates a new description. The domain specification defines the abstract description as an initially empty sequence that stores integer elements. The pre-conditions for push and pop specify that the sequence must not be full and empty, respectively, when the operations are performed. The post-conditions capture the result of performing push and pop as last-in-first-out append and extractions from the sequence, respectively. The concrete representation is an array of elements which the user of the abstraction can access only through the specified operations.

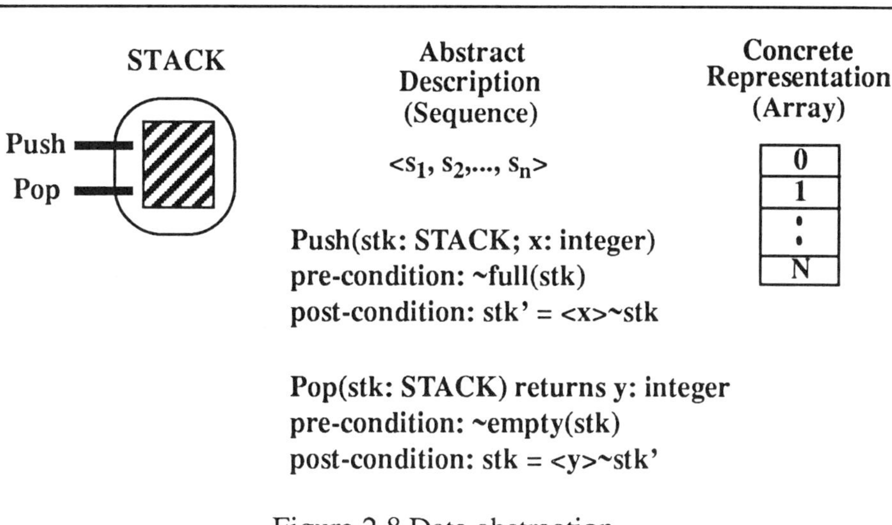

Figure 2.8 Data abstraction

Object-oriented design systems [133][134][135][136][137] employ data abstraction and produce software that is more malleable. In other words, the software supports change, reusability, and enhancement. Thus, the object-oriented design methodology helps to decrease costs during the design and operational phases, which is when the greatest software cost is incurred [9]. Some methodologies have tried to merge the structured and object-oriented views of design [101][138][139].

Software Synthesis. *Software synthesis* is the derivation of correct and efficient programs from higher-level specifications and reusable components [140][141]. The use of software synthesis capabilities has the potential of increasing software productivity, lowering development cost, and providing increased confidence that a given software implementation satisfies its specification. Some possible software specification techniques include natural languages, special purpose languages (for example, state transition diagrams), real examples (input/output pairs), logical formalisms (such as predicate calculus), and very high level languages (VHLL) [142]. VHLLs, such as SETL, Refine, and

Gist, provide features that allow programmers to ignore (defer) implementation details.

The roots of software synthesis date back to notions of automatic programming in the 1950's [142]. In those days, early compilers and assemblers were a significant step in software automation. Today's software synthesis systems strive to address higher levels of abstraction and multiple phases of the software life cycle: requirements, design, implementation, verification, and maintenance. These systems employ concepts and techniques from software engineering and artificial intelligence.

Several techniques are employed in software synthesis [142]. Transformational methods apply a sequence of transformations to the specification, which is gradually converted into a low-level implementation. Typically, the specification is in the form of a VHLL, and the transformations are correctness preserving. In deductive techniques, software synthesis is regarded as a theorem-proving task. Thus, inference systems are utilized which apply rules to a given set of initial facts to arrive at a goal fact. The simplest synthesis techniques are procedural methods. These techniques involve writing a special purpose program which generates the desired results. Current compilers and program generators typify this technique. Inspection techniques utilize knowledge of cliches, combinations of elements corresponding to familiar concepts. The aspect of "inspection" refers to recognizing cliches in specifications and selecting from a collection of cliched implementations.

Approaches to software synthesis can be classified as generic, application specific, and combinations of the previous two [140]. Generic approaches characterize common facets of software, such as data structures, algorithms, and data/control flow. These elements are common to all software regardless of the application domain. Application-specific approaches relate knowledge about the domain to various phases of the software life cycle. In this approach, domain-specific knowledge is represented in the form of rules.

Several systems are emerging that support software synthesis. Systems have been described which use hierarchical finite state machines for system specification and support algorithm design, data-type refinement, and validation [143]. The Sinapse system [144] is a knowledge-based, domain-specific system which synthesizes mathematical-modeling software. Model-based synthesis approaches [145] for reactive systems have also been presented. These approaches express domain knowledge as formal models.

2.5.3 Hardware Development

In this section, an overview of the hardware development process is provided. Next, the abstractions used are considered. Finally, hardware synthesis is discussed since these capabilities are becoming increasingly important within the design process and are being used as a basis for some codesign approaches.

Overview. Figure 2.9 illustrates the major phases involved in hardware development [3][6]. These phases include requirements, design, fabrication, and testing. This description is similar to the software development process, and many of the techniques described earlier can be applied to hardware development.

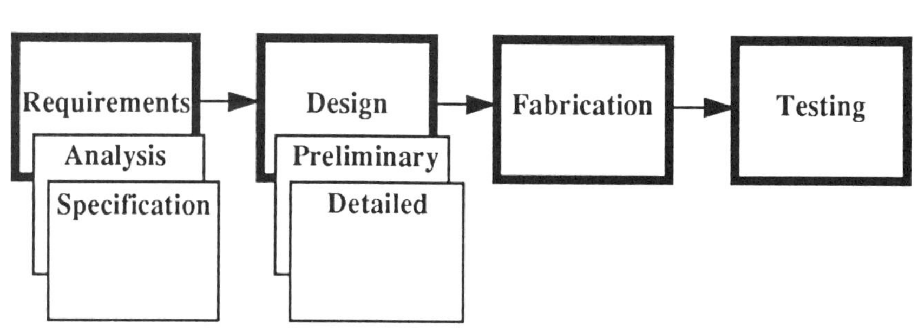

Figure 2.9 Major phases of hardware development

In addition to design constraints, for example, form factor and performance, the requirements phase involves understanding software considerations, such as requirements on the operating system. Environmental issues [146], such as whether the system is being developed for a military or a commercial application, are also important. Other factors include the available semiconductor technology.

Preliminary design establishes the functional modules necessary to implement the system and their interfaces. Available off-the-shelf components may be identified. Also, specifications are generated for the individual modules. Simulation and modeling can be used to further explore design options, evaluate candidate approaches, and establish constraints for individual modules.

Detailed design involves developing the internal logic for each of the individual modules. A module may be one or more boards or a single integrated circuit. Implementations can utilize off-the-shelf components or full custom approaches. More recently, programmable devices, such as field programmable gate arrays (FPGAs) [147], have become popular. Special purpose hardware is often implemented using FPGAs.

During this phase, behavioral models using hardware description languages can be developed to assess functionality and performance. In some circumstances, hardware synthesis (described below) can be utilized to generate an implementation. In other cases, additional refinement of the models may be necessary. In the development of microcoded machines, firmware design may be required [6]. This activity involves microprogramming.

In the fabrication phase, the physical implementation of the hardware is derived. This requires manufacturing boards, consisting of integrated circuits and other hardware components. The integrated circuits may be existing, off-the-shelf components or may themselves have to be manufactured. Finally, the hardware is tested. As in software development, several stages of testing are employed.

Abstractions. As indicated in the model of design representation provided in Table 2.1, hardware components can be described at different levels of abstraction. Most hardware description languages employ functional abstractions as a means of modeling these components. An example of an arithmetic logic unit (ALU) represented as a functional abstraction is depicted in Figure 2.10. In the same way as the "square_root" example of Figure 2.7, the ALU is treated as a "black box" with well defined inputs and outputs.

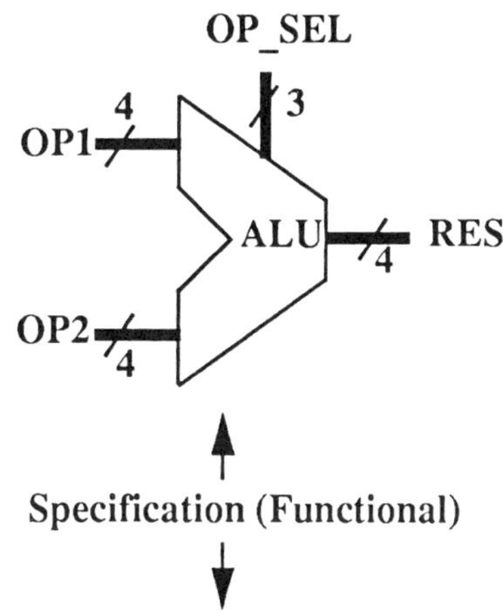


```
ENTITY ALU is
    PORT (OP1, OP2: IN BIT_VECTOR(3 DOWNTO 0);
          OP_SEL: IN BIT_VECTOR(2 DOWNTO 0);
          RES: OUT BIT_VECTOR(3 DOWNTO 0));
END ALU;
```

Figure 2.10 Functional abstraction (ALU)

At a more abstract level, hardware components can be considered resources. For example, at the system level, processors may be modeled as delay elements with queues. The concept of abstract resources and the use of data abstraction for modeling hardware components will be explored further in subsequent chapters.

High-Level Hardware Synthesis. *High-level hardware synthesis* (also called *behavioral synthesis*) capabilities [12] accept a behavioral description of a circuit along with a set of constraints specified in terms of speed, area, or power consumption. Constraints may also restrict the number of available hardware resources. From this information, synthesis tools generate a hardware structure which implements the behavior and satisfies the constraints. These synthesis systems require functional information and share many aspects with software compilers. Other tools can then be used to generate a physical implementation, such as an integrated circuit. Thus, synthesis capabilities, coupled with other tools, address the algorithmic, register-transfer, logic, and circuit levels of the design process.

There are many reasons for using hardware synthesis within the design process [12][148][149]. Many of the complexities associated with the design process can be hidden. Thus, one can focus on the behavior of a system without worrying about the details associated with an implementation. Also, the design space can be explored for alternatives and various trade-offs can be analyzed, such as performance versus area [12][150][151].

A typical high-level synthesis design process starts with a program-like behavioral description at the algorithmic level along with some constraints [12][152]. The description is converted into an internal representation, usually in the form of data/control flow although parse trees have also been utilized [153]. In the graph, the nodes correspond to operations, such as add and multiply. Some internal representations combine the data and control flow while others separate them.

Regardless of the form, the internal representation is the basis for performing compiler-like behavioral transformations, partitioning the behavior across multiple physical units, for example, several integrated circuits, and then performing control step scheduling and data path allocation. Given a data/control flow graph, control step scheduling assigns operations in the graph to controls steps, which represent clock cycles. Data path allocation provides functional units to implement the operations, memory to store data values, and interconnects (multiplexers, busses) to route the data. From the data path and schedule, a state machine is developed to control the sequencing of the elements in the data path. The end result is a register-transfer level description which can be given to module generation and logic synthesis tools [154][155] to generate a physical realization.

Control step scheduling and data path allocation are interdependent tasks. For example, it may be possible to perform certain operations in the same control step, requiring the use of multiple functional units. However, resource constraints may not allow this parallelism to be exploited.

A portion of a data flow graph undergoing control step scheduling and data path allocation is illustrated in Figure 2.11. For simplicity, any necessary multiplexers and busses are not shown in the data path. Because of the parallelism present, the graph can be implemented using one adder and one multiplier (along with other miscellaneous components) for which the total execution time would be three clock cycles (assuming the inputs have been loaded into registers). An implementation containing a single functional unit that performs both multiply and add would require four clock cycles. Other considerations include the minimization of interconnects and the number of registers.

Numerous synthesis systems employing a variety of approaches have been developed. Knowledge-based approaches, such as DAA [156] and HAL [157], have been described. SEHWA [158] has been used to synthesize pipelined architectures. MIMOLA [159] synthesizes the data path and control for CPU structures. Other synthesis capabilities include CAMAD, Chippe, and Emerald, described in the Carnegie-Mellon University research effort [152], and USC's ADAM [160].

Synthesis systems have been used successfully for special purpose processors, particularly digital signal processors [161][162][163][164]. The need for considering low-level information to improve the overall synthesis process has been recognized [151][165][166]. Also, several approaches for performing hardware partitioning have been developed, allowing a behavior to be implemented using multiple integrated circuits [167][168][169].

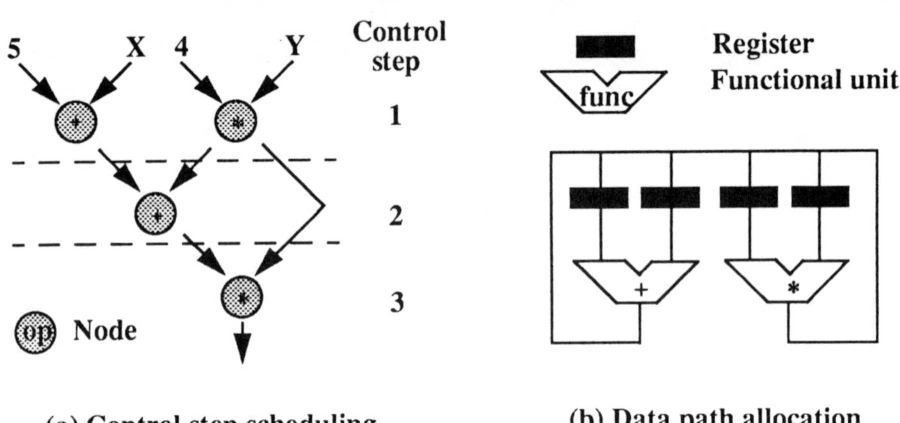

Figure 2.11 Control step scheduling and data path allocation

2.6 SUMMARY

This chapter has presented an overview of several topics that are important to hardware/software codesign. Embedded systems were introduced. These systems contain a mixture of hardware and software tailored for a particular application. Codesign can benefit the development of such systems. The sections regarding models of design representation and the virtual machine hierarchy provided a framework for discussions on hardware/software development and abstractions. Several performance modeling techniques and environments were mentioned which could be employed to analyze hardware/software systems early in the design process. The development process for hardware and software was described, along with an emerging technology, that of synthesis.

Some themes of particular relevance to the monograph have also appeared in this chapter. One theme is complexity management. Complexity considerations are common to both hardware and software. These considerations are addressed through the use of abstractions by focusing on only those aspects of interest.

Concepts and technologies were presented that suggest possible approaches for unifying the hardware and software design processes. Foremost, data/control flow concepts are prevalent in both domains. As a result, data/control flow representations can serve as a unified representation for both hardware and software. In addition, the discussion on virtual machine abstractions implies that there is little difference between hardware and software. The only difference is the level at which hardware and software appear within the virtual machine hierarchy.

Chapter 3
Hardware/Software Codesign Research

Having taken a look at some important aspects of hardware and software development, it is now possible to embark on a discussion of hardware/software codesign. Attention is given to both past and current research. This chapter starts with an informal view of codesign and presents two related areas: hardware/software trade-offs and hardware/software cross fertilization. Next, a description of a typical codesign process is provided, and several topics are discussed within the context of this design process. Some codesign approaches and design environments are then presented. Following this discussion, a unified design environment is briefly described. This environment will be used to demonstrate many of the ideas developed in the monograph.

3.1 AN INFORMAL VIEW OF CODESIGN

It is natural to pose the question, "What does the *Co* in codesign mean?" The two most common interpretations are concurrent (design) and cooperative (design). The term concurrent also has two possible meanings. The first is the design of the hardware and software in the form of parallel (independent) paths, which constitutes the current view of the design process. The second refers to the joint (combined or united) development of the hardware and software. The latter captures the notion of "interaction" between hardware and software design, whereas the former does not. A stronger expression of this interaction is conveyed by the term cooperative. Although the parallel development of hardware and software is referred to as codesign by some individuals, the aspect of interaction is essential. Thus, the term cooperative and the second interpretation of concurrent should be kept in mind when discussing codesign in this monograph. These ideas are illustrated in Figure 3.1.

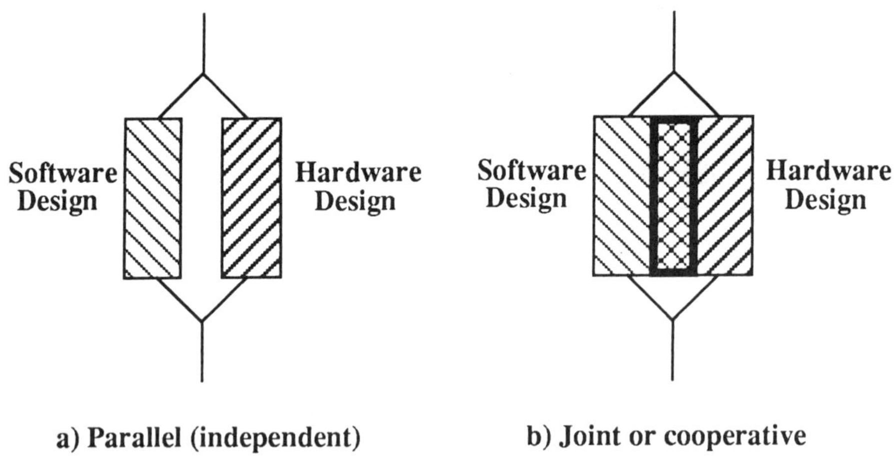

Figure 3.1 Parallel design versus joint or cooperative design

Interactions between software and hardware design occur in different ways and at various levels. The ISA is typically considered the boundary or interface between software and hardware [170]. As a result, design interactions occur between compiler writers and hardware developers as functionality is transferred across this boundary.

Architectural trade-offs are involved in deriving a suitable instruction set and include exploring the use of various hardware features, such as caches and pipelining, to support a given algorithm [171]. Another example of interaction is the interfacing of a software program to a hardware device, such as a disk controller or an ASIC. In parallel and distributed systems, interactions take place during the partitioning/ scheduling of software across multiple processors. This design activity encompasses the evaluation of different interprocessor communication (IPC) strategies.

3.2 HARDWARE/SOFTWARE TRADE-OFFS

The process of performing hardware/software trade-offs requires a direct interaction between hardware and software design. *Hardware/ software trade-offs* refer to decisions regarding the allocation of functions into hardware and software that attempt to satisfy a set of objectives. Thus, as shown in Figure 3.2, hardware/software trade-offs affect the amount (and mixture) of software and hardware employed, and lead to other kinds of trade-offs, such as performance versus cost [172].

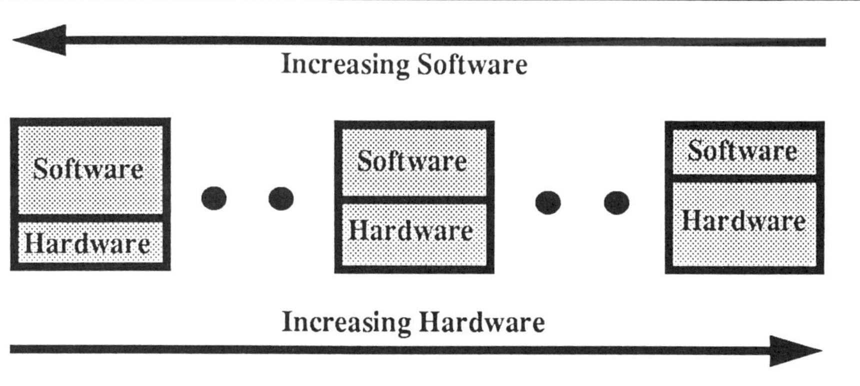

Figure 3.2 Impact of hardware/software trade-offs

Hardware/software trade-offs are not new to the design of complex systems. These trade-offs have been described in the literature within the context of fault tolerant computers [173], buffer memories [172], minicomputer design [174], pipeline interlocks [175], and real-time systems [176]. Many trade-offs occur during instruction set design [1][174][177]. Some trade-offs involve the migration of software functions into firmware [178][179]. This approach has been utilized to increase the performance of operating systems [177][179]. Other trade-offs involve performing specific tasks in dedicated coprocessors [180]. High-level hardware synthesis capabilities extend the range of hardware/software trade-offs that can be explored by allowing functions to be implemented as special purpose hardware.

3.3 CROSS FERTILIZATION

The *cross fertilization* between the software and hardware domains (also referred to as technology transfer) [181][182] is particularly relevant to codesign. As indicated in Figure 3.3, cross fertilization refers to the application of software engineering concepts to hardware engineering (and vice-versa). This area addresses such issues as the similarities and differences between software and hardware, as well as the unification of the software and hardware development paths. The latter includes the investigation of unified representations for hardware and software, a topic that is explored further in this chapter.

3.3.1 Similarities and Differences Between Software and Hardware

Before focusing on explicit cross fertilization and techniques for unifying the two design paths, it is worthwhile to discuss some of the similarities and differences between hardware and software [181]. From a design process point of view, complexity management is critical in both domains, and abstraction is an important tool in dealing with complexity. Also, the ability to communicate change and catch errors, such as inappropriate design decisions, early in the design process is crucial.

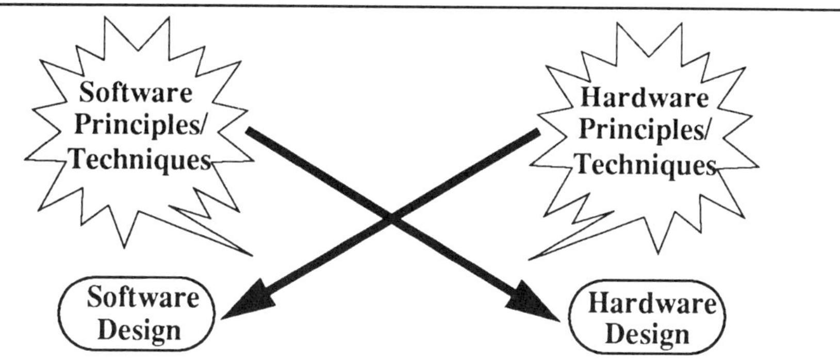

Figure 3.3 Cross fertilization between software and hardware design

Many concepts are familiar to both hardware and software domains. Modules and reusable elements are employed in these domains, although reusable elements are better established in the hardware domain. For example, consider the libraries of small scale, medium scale, large scale, and very large scale building blocks that are available for hardware design. In addition, hardware and software modules share common characteristics. Namely, the notions of function, state, and programmability apply to both.

The most glaring difference between hardware and software is that hardware has physical characteristics, such as delay and power consumption, while software does not. As a result, there is no counterpart in software to the fabrication of integrated circuits that occurs in hardware. Because hardware has physical constraints, it can be argued that aspects of hardware design are easier since the number of design options is restricted. However, these physical characteristics also make hardware design difficult. Timing considerations, noise effects, heat dissipation, and electromagnetic effects complicate hardware design and, in many cases, prevent correct operation.

Although there are similarities between the two design processes, there are some differences as well. Hardware designers make extensive use of visualization environments and simulation models before committing to a particular design. For example, schematic capture facilities are used to construct and visualize designs. Once captured, these designs can be simulated to determine their performance and correctness. Although there is renewed interest in visualization environments for software, simulation and modeling are not as widely used in the software design process. Another difference is the emphasis on functional correctness in software design. Although hardware designers also seek correct functionality, performance requirements currently drive many design decisions, such as those that pertain to space-time trade-offs.

3.3.2 Explicit Cross Fertilization

Cross fertilization has occurred in both directions [181]. Some examples of cross fertilization from VLSI design to software design include the use of graphics-driven, visualization environments and the use of executable models to evaluate systems earlier in the design process. Several individuals have used flowchart-like descriptions to represent software. However, these descriptions become complicated and clumsy when designing software of even moderate complexity [183]. Graph-based approaches have been utilized more effectively for analyzing software.

Examples of cross fertilization from software engineering to VLSI design include the use of functional abstractions and hierarchical decompositions in the design process. Another example is the application of compiler-like transformations and techniques to high-level hardware synthesis capabilities, such as dead code elimination, loop unrolling, and register allocation using lifetime analysis. Some efforts have tried to address design change management issues. Concepts such as information hiding, program families, and the use of abstract interfaces have been employed in an attempt to minimize the impact of change in the VLSI design process.

3.3.3 Unifying the Software and Hardware Development Paths

In an effort to merge the hardware and software development paths, the use of uniform representations and design techniques for both hardware and software has been attempted in various forms. One of the earliest works in this area was by Wulf [184]. An important idea in that work was the emphasis on the functional aspects of a system before considering implementation in hardware or software. The research led to a notation that could be used to describe hardware or software.

The EPOS environment used a uniform methodology, consisting of hierarchical design techniques for both hardware and software based on functional abstractions [185]. Once the system functions to be considered for implementation were defined, decisions were made regarding whether the functions should be implemented in hardware or in software. Functions that were implemented in software would require preliminary and detailed design.

The SARA environment [75], a multi-level design system, is utilized for concurrent hardware or concurrent software design. A language called SL defines a structural hierarchy constructed out of modules, sockets, and interconnections. Another language called the Graph Model of Behavior (GMB) is used to describe the system's behavior.

The GMB consists of separate data and control flow and can be embedded within the modules of the structural description. This representation uses constructs from three domains: control, data, and interpretation. The control domain consists of nodes and control arcs which are responsible for the sequencing of the data flow. This domain is similar to Petri nets. The data domain consists of processors (also called transform units), data sets (data values), and data arcs. The interpretation domain defines the data types of the data sets and the algorithm for the various processors. PL1 is used to describe the behavior of the processors. It is possible to simulate the GMB using a graph constructed from only the control and data domains. This aspect is similar to the concept of uninterpreted modeling mentioned earlier.

Constructs in the data flow can represent software or hardware elements. Thus, hardware/software trade-offs can be performed. However, one limitation with this environment is that multiple representations are required, and behavioral descriptions necessitate the separation of data and control flow [22].

3.4 A Typical Codesign Process

Figure 3.4 describes a methodology commonly employed in hardware/software codesign approaches today. The codesign process starts with a functional, system representation independent of hardware and software. Several system representations may be utilized, including finite state machines (FSM) and concurrent processes. In some circumstances, the system is described using a programming language which is then compiled into an internal representation, such as a data/ control flow description. This description serves as a *unified representation*, one that can represent software or hardware.

Hardware/software partitioning, the process of determining which functions are to be provided in hardware and which in software, is performed on the unified representation. This representation can be an internal description or the system description itself. The partitioning can be accomplished manually or automatically. The result is a *hardware/ software partition*, which specifies the functions to be implemented in hardware and in software. From this description, the hardware and software for the system are generated. Typically, hardware synthesis and software synthesis techniques are employed, respectively. Any interfaces required between communicating software (plus processor) and special purpose hardware components are also synthesized.

The goodness of the partitioning decision can be evaluated at two different points, designated as (A) and (B) in Figure 3.4. In some approaches (A), estimation techniques are used to evaluate the hardware/software partition, typically in terms of performance. If the evaluation does not meet the required objectives, another hardware/software partition is generated. In other approaches (**B**), a hardware/software implementation derived through synthesis is evaluated with respect to various metrics at the instruction set level. If the result of the evaluation is unsatisfactory, another hardware/software partition is created. These steps are performed iteratively until a satisfactory hardware/software implementation for the system is developed.

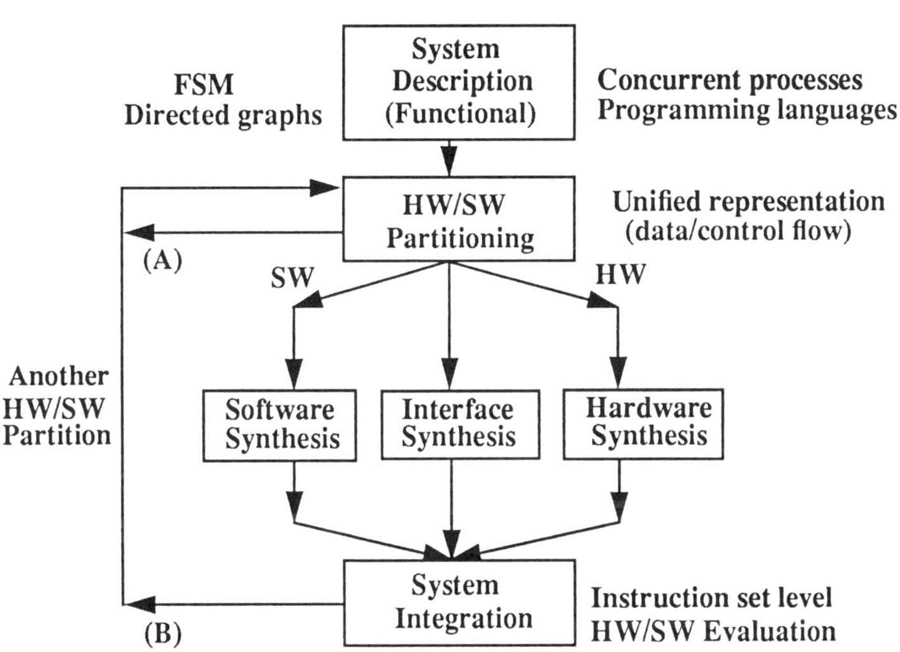

Figure 3.4 A typical codesign process

The most notable differences between this codesign process and the current design process (see Figure 1.1) include the use of a unified representation for describing hardware and software, and the ability to perform hardware/software partitioning iteratively. Within the context of this codesign process, related work spans several areas. These areas include the use of unified representations, hardware/software partitioning approaches, and metrics for evaluating hardware/software systems.

3.4.1 Unified Representations

Within a codesign process, unified representations play an important role in two primary instances. Before hardware/software partitioning is performed, it is necessary to have a description that can be implemented in either hardware or software. Such a description allows systems to be viewed independently of hardware or software, supports a delayed binding hardware/software partitioning strategy [186], and allows synthesis techniques to be employed. The delayed binding approach to hardware/software partitioning is discussed in the next subsection.

After hardware/software partitioning, a unified representation plays an integral role during the hardware/software evaluation and refinement stages. Specifically, a unified representation supports the early evaluation of hardware/software systems in a common simulation environment and thus, enhances the communication between hardware/software developers through the use of a common modeling paradigm. Hardware/software trade-offs can be performed more easily by allowing functionality to be transferred between software and hardware. A unified representation also allows the possibility of cross fertilization between the software and hardware domains. Therefore, techniques and results from one domain can be applied to the other.

Graph-based representations, intermediate representations in compilers, finite state machines (FSMs), and concurrent processes provide unified representations for both hardware and software. Graph-based representations, such as data flow graphs and Petri nets, can be used to represent hardware or software. For example, nodes in data flow graphs can represent operations in software or functional units in hardware. Graph-based representations have a natural ability to model data/control flow and concurrency. Intermediate representations in compilers [187], such as abstract syntax trees and directed acyclic graphs, are used to support automated synthesis.

FSMs can be used for expressing system behavior, hardware modeling, and software modeling. Communicating FSMs are useful for describing reactive real-time systems [188]. FSMs provide a unified, mathematical foundation which can be utilized to perform formal verification of correctness, simulation, hardware/software partitioning, and synthesis. Several individuals have noted the similarity between FSMs and object-oriented approaches [101][189][190]. In particular, objects can be viewed as communicating finite state machines with each object representing a machine whose state can be manipulated via methods [189][190]. Other system representations include the use of statecharts [109], which are extensions of FSMs.

Concurrent processes [23][65][191], which are fundamental to the modeling of both software and hardware, can also serve as a unified representation. This representation has a strong foundation from works such as communicating sequential processes (CSP) [192]. Also, several well defined concepts already exist [23].

Some efforts in codesign have attempted to "map" several modeling paradigms onto an intermediate modeling representation. This multiparadigm mapping approach has utilized FSMs [188], intermediate models [193], and integrated modeling substrates [194][195] as the common, underlying representation. Most of these approaches are at the conceptual stages of development.

An integrated modeling substrate is shown in Figure 3.5. The idea is to allow existing, familiar paradigms in hardware and in software (VHDL, Verilog, ADA) to be mapped onto this substrate, which provides a unified, internal representation. The intent is to support a wide range of hardware/software trade-offs, permit continuous and incremental hardware/software integration, allow more alternatives to be explored, improve model continuity, and support early evaluation.

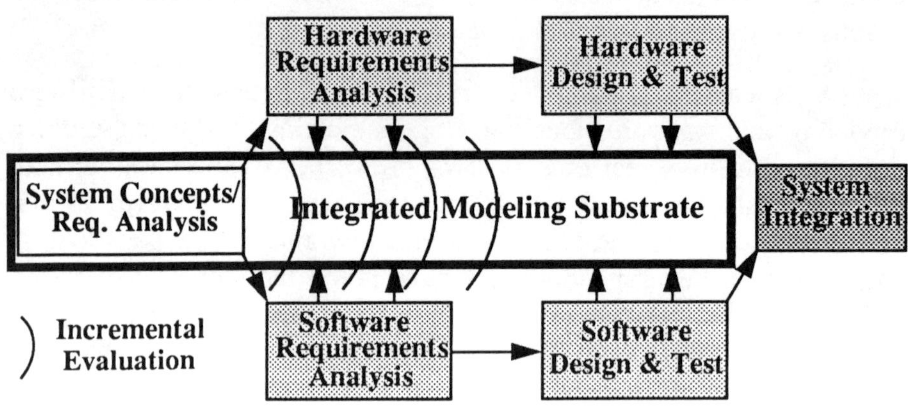

Figure 3.5 An integrated modeling substrate (Modified from [195], © 1992 IEEE)

Some work is also underway in the area of developing specification languages. Examples include the use of object-oriented specification languages [196] and specification languages for composite systems [197]. Specification techniques provide a common framework for representing software or hardware.

3.4.2 Hardware/Software Partitioning Approaches

Hardware/software partitioning approaches can be classified with respect to various attributes [198]. These attributes include the type of representation used as a starting point for partitioning, the partitioning algorithm, the amount of automation, the granularity of the functions being considered, and the design stage at which partitioning takes place. The initial specification can be a hardware specification, a software specification, or a more general specification that is independent of hardware or software. The partitioning algorithm determines which nodes are to be implemented in hardware and which in software. Cost functions are used to evaluate the quality of the partitioning decision. Numerous, well known algorithms can be applied here, such as greedy algorithms and simulated annealing. The degree of automation ranges from manual to fully automatic. The automated approaches are typically iterative, starting with some initial description and gradually "migrating" to a hardware/software implementation that satisfies the objectives.

The partitioning granularity, which refers to the size of the function (unit) considered for allocation into hardware or software, can be coarse-grain or fine-grain [199]. Coarse-grain partitioning utilizes tasks or functions, and fine-grain partitioning employs basic blocks or individual statements. The concept of virtual instruction sets is closely related to partitioning granularity. This idea will be explored further in the next chapter.

Regarding design stage, an early binding or a delayed (late) binding partitioning strategy [186] can be used. An early binding strategy commits functions into hardware and software at the beginning of the design process, while a delayed binding strategy defers the hardware/software partitioning decision. The latter is desirable in situations where it is difficult to ascertain whether a function should be implemented in hardware or in software. A discussion of the implications of these strategies will also be provided in the next chapter.

Two common types of hardware/software partitioning approaches have been described: hardware-oriented [200] and software-oriented [25][201]. In a hardware-oriented approach, an initial hardware/software partition is selected, consisting predominantly of functions implemented in hardware. Functions implemented in hardware are gradually extracted and moved to software. This process continues until performance constraints are violated.

A software-oriented partitioning approach starts with an all software solution, gradually extracts performance critical portions, and implements these portions in hardware. Simulation or program profiling can be used to determine these bottlenecks. In these approaches, the 90/10 rule for programs often applies, where 90 percent of the time is spent in 10 percent of the code. Thus, time-consuming loops are often candidates for movement into hardware. This approach is useful for "instruction set metamorphosis" [202], the augmentation of a core processor's functionality with new operations.

Other hardware/software partitioning approaches use clustering [203] to control the process of hardware/software partitioning. Some approaches [198] perform both coarse-grain and fine-grain partitioning in successive steps. Instead of using a single objective function, the global criticality local phase (GCLP) algorithm [204] selects an appropriate objective function at each step. The selection of an objective function is based on (global) time criticality and local node characteristics, which reflect tendencies towards software or hardware implementation.

3.4.3 Metrics for Evaluating Hardware/Software Systems

Several metrics exist [205][206] for parallel systems that are applicable to hardware/software evaluation. Some of the metrics include execution time, speedup, efficiency, overhead ratio, utilization, redundancy, and price. Other examples are the operator sensitivity metric for software performance [77] and the module sensitivity metric for reliability [85]. The operator sensitivity metric identifies bottlenecks in software by quantifying the portion of execution time taken by a software node. Software nodes with high sensitivity values are good candidates for replacement by hardware.

The module sensitivity metric for reliability assesses the impact of changing the reliability of an individual software module on the overall system software reliability. A module with a high sensitivity indicates that the module has a great impact on overall reliability. Thus, the metric suggests that it is worthwhile to improve those modules with high sensitivity values. This approach utilizes Markov models to determine the metric.

Several cost functions, many of which are used by automated hardware/software partitioning approaches, have been developed to evaluate hardware/software partitions. The cost function in the Cosyma system [199] takes into account communication overhead between a core processor and a special purpose processor as well as parallel execution between the two processors. The Stanford cosynthesis system [200][207] employs a cost function consisting of a weighted sum of hardware size, software size (program and data), bus utilization, processor utilization, and communication overhead.

3.5 CODESIGN ENVIRONMENTS

This section presents an overview of several hardware/software codesign environments. These environments are classified with respect to some distinguishing characteristic or property. The intent of this discussion is to convey the broad range of approaches currently being investigated.

3.5.1 Graph-based Software/Hardware Mapping Approaches

Several environments have been developed to support performance analysis early in the design process using abstract hardware/software models. Many of these approaches are based on the use of some kind of directed graph representation, such as data flow graphs, as a means of representing software and hardware. This graph representation serves as a unified representation. In order to analyze the performance of a system, the software representation is mapped onto the hardware representation.

As illustrations of this type of environment, the ADAS [208] and Dproto [209] modeling environments employ software graphs and hardware graphs to describe software and hardware, respectively. A mapping assigns the software operations to the hardware processing elements. Upon accomplishing this mapping, the resulting performance is assessed.

In ADAS, this mapping is represented via the projected processing graph (PPG), which is a constrained software graph. User defined priorities are employed to resolve contention for hardware resources by software operations. Similar to processes within an operating system, software operations that contend for a hardware resource cycle through several states, such as ready and blocked. When software operations are enabled, they execute for a time specified by a delay.

Two kinds of analysis techniques are supported in ADAS: analytic and simulation. Analytic techniques [63] employ Molloy's stochastic Petri net results [55] to determine various performance metrics, such as the utilization of the hardware. Two levels of simulation are supported. A coarse grain simulation is used to allow performance analysis without considering the detailed functionality of the nodes or the contents of the tokens.

In a functional simulation, the software operations correspond to descriptions in "C" or Ada. The tokens that circulate in the software graph carry data structures with actual values. Thus, this level of simulation is used to verify functionality. Once the functionality and performance of a design has been assessed, VHDL can be generated for the hardware descriptions.

3.5.2 Rapid Prototyping Approaches

Codesign can be used for the rapid prototyping of systems. Rapid prototyping attempts to shorten the design path from a system specification to a prototype through improved electronic design automation techniques, such as automated synthesis. This approach is useful for requirements validation, testing various properties of a design, and shortening the design time of a system.

The SIERA [210] design environment focuses on the rapid prototyping of application specific systems. Because of the emphasis on rapid prototyping, the approach utilizes architecture templates and library modules. This environment uses VHDL for system specification. The system specification consists of a hierarchical set of concurrent, sequential processes that interact via a communication mechanism based on first-in-first-out (FIFO) queues. Performance analysis can be carried out on the system specification. The use of concurrent processes provides an abstraction for both hardware and software processes, allowing different hardware/software allocations to be evaluated.

An architecture template serves as the base hardware/software architecture for the system to be designed. This template is a distributed structure consisting of single board computers and custom boards. These boards run a real-time operating system. The custom boards consist of processor modules and application-specific slaves. A manual mapping of software processes and hardware processes to the architecture template is performed. Software processes can be mapped to processors on any of the boards. Also, several processes can be mapped to a single processor. Hardware processes are mapped into application-specific slaves which interface through a device driver that executes on a programmable processor.

The library consists of hardware modules and programmable processor/kernel modules. Examples of hardware modules include parameterizable multiprocessors, uniprocessors, and memory. Digital signal processors and their associated kernels are also available. Various hardware and software generators have been constructed to allow the creation of hardware and software.

3.5.3 Multiparadigm Approaches

The PTOLEMY design environment [211][212] is a multiparadigm simulation approach which supports the co-existence of different models of computation. In PTOLEMY, a model of computation is called a domain. More precisely, a domain realizes a model of computation and contains, among other things, a collection of executable elements called blocks whose execution order is determined by a scheduler. Domains can be mixed together to describe an entire application through the use of interfaces called wormholes.

Some of the more prominent domains that are supported include synchronous data flow (SDF), dynamic data flow (DDF), discrete event (DE), and digital hardware modeling. The SDF domain consists of data driven blocks whose firing order is determined statically (at compile-time). In a DDF domain, the scheduling of blocks occurs dynamically (at run-time). The DE domain is a general environment for supporting time-related simulations, such as queueing-based simulation. The digital hardware modeling domain includes a functional simulator called THOR, which is used for gate-level and behavioral simulation.

Codesign starts with some algorithmic description. The algorithmic description is simulated to verify functionality. Hardware/software partitioning is performed manually (an automated approach is under development [204]). Next, a hardware simulation is performed, and using this hardware configuration, software is synthesized. The combined hardware/software is then simulated for functionality and performance. This process can be repeated iteratively.

PTOLEMY supports multiprocessor implementations. Relevant issues include determination of the number of processors, the IPC strategy, and software generation. The GABRIEL subsystem is used for software generation. GABRIEL takes the number of processors, the processor connectivity, and the IPC strategy into account and performs code partitioning to match the available hardware.

3.5.4 Cosynthesis Environments

Because embedded systems can be thought of as a generalization of ASICs [26], hardware synthesis systems are being extended to allow for *cosynthesis*, the generation of both the hardware and software components of a system. Most cosynthesis systems use data/control flow as the internal representation. Note that these approaches can be employed as part of a rapid prototyping environment. Efforts have focused on developing the hardware/software communications interface between a microcontroller and hardware devices [213], developing hardware/software partitioning algorithms [199][207], and synthesizing the hardware and software for a portion of functionality [214][215][216][217].

As an example, the Stanford cosynthesis system [216][217] accepts a behavioral description along with performance constraints as input.

The goal is to arrive at a hardware/software implementation which satisfies the desired behavior and the performance constraints. The performance constraints are in the form of latencies (the time to perform a task) or input/output data rates, which reflect the consumption/production rates for a task. The target architecture consists of a single, off-the-shelf microprocessor, one or more ASICs, and a single-level memory.

The system to be implemented is described using a language called HardwareC, a subset of "C", which is then converted into a system graph model, an internal data/control flow representation of system behavior. A tool called Vulcan II is responsible for performing a hardware-oriented hardware/software partitioning. Operations with unbounded delay are referred to as nondeterministic delay (ND) operations. ND operations can be internal, such as data-dependent loops, or external, for example, input/output operations. The remaining operations are considered deterministic (known at compile-time) since their delay is bounded. An initial hardware/software partition is chosen in which all deterministic and external ND operations are provided in hardware, while internal ND operations are implemented in software.

Starting with an initial hardware/software partition, new hardware/software partitions are created iteratively. Using a greedy algorithm, nodes are moved from hardware to software, taking into account communication costs and checking timing constraints. Only the deterministic operations (nodes) are considered for movement. In other cosynthesis systems [198][199], simulated annealing has been employed.

A cost function is used to evaluate the resulting hardware/software partition. The cost function is a weighted combination of hardware size, software size, bus utilization, processor utilization, and communication overhead. The partitioning produces a set of concurrently executing hardware and software models, which are also graphs, and an interface graph model used to interface the two models (see Figure 3.4). These graphs are then synthesized into a final hardware/software implementation.

3.5.5 Cosimulation Approaches

Cosimulation approaches [218][219][220][221][222][223][224] address the separation between the software and hardware design processes by allowing hardware/software simulation (integration) to occur earlier in the design process. Some systems focus on integration at the instruction set level [225][226]. Because hardware models are employed in the simulation, these techniques support the functional validation of the software before hardware is available. Cosimulation is especially applicable in large software-dominated systems, those with substantial amounts of software.

Most approaches link a software run-time environment with a hardware simulation. As a result, familiar hardware (VHDL, Verilog) and software environments (C, C++) can be used. The software and hardware execute as separate processes and communicate through UNIX IPC mechanisms, such as sockets. Thus, the software and hardware processes can execute on distinct hosts, communicating across some local area network.

For example, the Verilog simulation environment has been employed for supporting cosimulation. In such a scheme, Verilog is used to model hardware devices, such as processors and ASICs. A programming language interface, consisting of a collection of routines, is utilized to access information in the hardware simulation. In this environment, a set of calling functions from within the software application can be employed to access individual registers within the hardware simulation. Interactions from the hardware simulation to the software application, such as processing of interrupts, can also be supported.

3.5.6 Miscellaneous Codesign Environments and Efforts

The CODES environment [227] is based on the use of parallel random access machines (PRAM) as a system abstraction for a hardware/software system. To support the PRAM model, statecharts and tools based on CSP are utilized during the requirements engineering phase. After performing hardware/software partitioning, VHDL and "C" are used to support the hardware and software design processes, respectively. The goal of this environment is to support the current design process using existing tools and techniques.

There are several other efforts that have been described within the context of codesign. Environments have been developed which focus on real-time evaluation using rate monotonic analysis (RMA) [228][229], as well as the analysis of distributed systems [230]. Some work has addressed instruction set level interactions, such as code reorganization techniques for minimizing pipeline interlocks [175][231] and increasing the hit ratio of caches for particular applications [232]. Another codesign effort is the generation of compilers and simulators from an instruction set architecture description [233]. This work captures the idea of "joint" hardware/software development.

SES/Workbench [76] is a commercial modeling and simulation environment that uses hierarchical data/control flow graphs, consisting of nodes and arcs, to model system behavior and analyze performance. The directed graphs can be used to represent both hardware and software. In addition, the data/control flow representation can be synthesized into a hardware realization using a structural translator [234]. Another commercial offering is the JRS IDAS toolset [235]. IDAS supports multiprocessor synthesis/performance analysis, application specific processor synthesis, and retargetable code generation.

3.6 Limitations of Existing Approaches

There are several aspects of a design environment that are necessary in order to support codesign. The design environment should permit early performance evaluation and trade-off exploration. Such an approach allows hardware/software partitioning decisions to be evaluated *before* committing to a particular design. Because of the complexity of today's systems, hardware/software abstractions are required to evaluate systems quickly using abstract models, as opposed to detailed, instruction set level models. Also, the environment should not force all portions of a system description to be provided at the same level of detail (such as the instruction set level) before allowing any evaluation. In other words, incremental evaluation should be supported, enabling one to focus on those aspects of interest while ignoring other, inessential details.

Many of the existing approaches to codesign violate the above tenets. For example, cosimulation focuses primarily on hardware/software integration. These approaches typically assume that several hardware and software design decisions have been made. Specifically, the software is fairly well developed, and in many circumstances, the hardware has already been selected. More importantly, the software and hardware are designed in separate environments. As a result, it is difficult to perform hardware/software trade-offs.

Cosynthesis environments also have limitations. Incremental evaluation using abstract models is not supported. In some approaches, hardware/software evaluation occurs at the instruction set level using very detailed models, only *after* synthesis has been performed (as in a typical hardware/software design process). This detail limits the number of design alternatives that can (or will) be explored and the types of systems which can be analyzed.

In general, current codesign environments focus purely on performance [14]. These environments do not evaluate systems with respect to multiple system level metrics, such as performance and reliability, which is important for many embedded systems. This capability is also important when one portion of the system requires high performance while another portion requires high reliability.

The critique provided in this section applies to many of the environments discussed in this chapter, with the exception of ADAS. However, one limitation with ADAS is the inability to smoothly incorporate lower level implementations as the model is refined. Thus, model continuity is not supported.

3.7 THE ADEPT MODELING ENVIRONMENT

ADEPT (Advanced Design Environment Prototyping Tool) [21][236][237] is a unified modeling environment which is being developed to support end-to-end system design. This environment will be used to demonstrate many of the concepts and ideas developed in this monograph. In ADEPT, models are constructed with a collection of high level, data/control flow modules [64]. These modules model the flow of information through a system via tokens and communicate using a uniform, asynchronous handshaking protocol. Each ADEPT module is described as a VHDL process and has a corresponding colored Petri net representation [238]. Higher level modules can be constructed from the basic set of library modules. In addition, custom modules can be incorporated into the design as long as the handshaking protocol is maintained.

Referring to Figure 3.6, a token in this environment is implemented as a VHDL record structure. In the token, the status field is used for performing handshaking between modules, and the color field is utilized for user-specified information. Library modules are provided which can manipulate the information in the color field. For example, modules exist that allow a user to perform functional transformations on the contents of the tokens.

An example of a library module is the *WYE*, shown with its underlying Petri net representation in Figure 3.6. When a token arrives at In1, tokens are placed simultaneously at Out1 and Out2. When the tokens at Out1 and Out2 have been acknowledged, the token at In1 is acknowledged.

In the Petri net representation, the handshaking is emulated via the ready ("r") and acknowledge ("a") places. When a token arrives at the place labeled "0r", the top transition is enabled, and a token is transferred to the "1r", "2r", and center places. Once the output tokens are acknowledged (tokens present at the "1a" and "2a" places), the lower transition is enabled, and a token is transferred to the place "0a", indicating that the input token has been acknowledged. The module is then ready for the next input token. Other modules are modeled similarly.

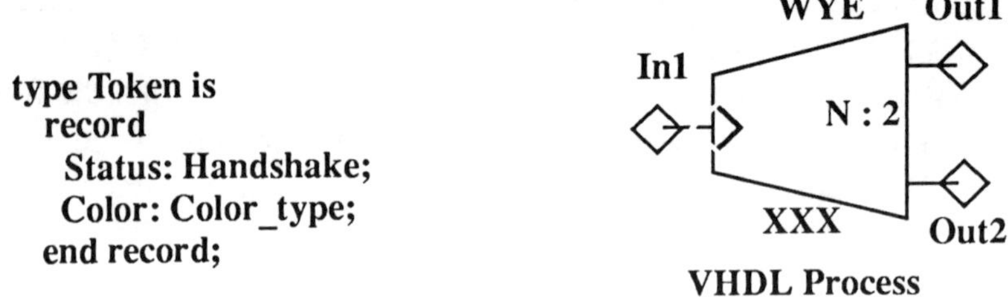

a) Token record structure b) *WYE* module symbol

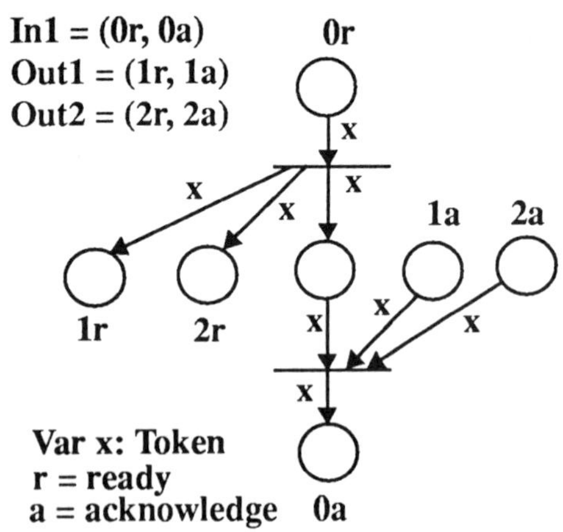

c) *WYE* CPN representation

Figure 3.6 Token definition, *WYE* module, and Petri net

There are several aspects of the ADEPT environment which make it promising for codesign exploration. The environment promotes the idea of unified representations, an important characteristic of codesign approaches, through the use of a common set of library modules which communicate in a uniform fashion. Because data/control flow concepts are utilized, the modules can provide a common modeling paradigm for hardware and software. In this environment, integrated performance and reliability are supported. From a codesign perspective, an integrated design environment is appealing since hardware and software descriptions can be evaluated collectively within a unifying framework.

3.8 SUMMARY

This chapter has presented an overview of codesign research. The chapter started by informally defining codesign as a cooperative design approach, stressing the interaction between software and hardware design. Two areas relevant to codesign were discussed: hardware/software trade-offs and the cross fertilization between software and hardware design. The presentation on cross fertilization described some of the similarities and differences between hardware and software. Also, some early codesign efforts which attempted to unify the hardware and software design processes were briefly mentioned.

An overview of a typical codesign process was provided, along with more detailed discussions on unified representations, hardware/software partitioning, and metrics for evaluating hardware/software systems. The latter included some commonly used metrics. In addition, various cost functions used in automated hardware/software partitioning approaches were discussed. To provide additional insight into the different approaches attempted, several codesign environments were introduced. The chapter concluded with some limitations of these design environments and a brief description of the ADEPT modeling environment.

Chapter 4
Codesign Concepts

In order to establish a common base for subsequent discussions of codesign, some definitions of relevant concepts, based on those found in [239], are presented in this chapter. In the notation that follows, the disjunction symbol "$\vee$" is read as "or". A set "$\{\ldots\}$" refers to a collection of elements, and the empty set "$\emptyset$" is a set with no elements. The union "$\cup$" of two sets consists of all elements belonging to at least one of the sets. The intersection "$\cap$" of two sets consists of all elements common to both sets.

4.1 Functions

Definition 4.1 *Function*: a mapping from inputs to outputs, which may be based on state.

The inclusion of state within the definition implies that successive invocations of a function with the same inputs may not necessarily produce the same output. When necessary, a subscript will be introduced in the notation for clarity. Thus, the symbol f_k will denote "function k", where k is an index or a qualifier used to distinguish between several functions. For example, f_2 is read as "function two", and f_{sqrt} corresponds to a square root function.

The notation shown in equation (4.1) will be used to express a mapping from a set of inputs X and state S to a set of outputs Y and a new state S'.

$$f: (X, S) \rightarrow (Y, S') \tag{4.1}$$

In some circumstances, it will be necessary to describe a mapping from one or more elements of X to one or more elements of Y (assuming no state). The notation in equation (4.2) will be used to express this mapping.

$$(y_1, y_2, \ldots, y_n) = f(x_1, x_2, \ldots, x_m) \tag{4.2}$$

For future reference, a *system function* will refer to a function to be performed by the system under consideration. Initially, system functions are independent of a hardware/software implementation. A *hardware/software implementation* is a particular mixture of hardware and software that is assigned to a system function. In general, the term *implementation* will refer to the realization of some description of behavior, such as a functional specification.

4.2 Functional Decomposition

Definition 4.2 *Functional decomposition*: the process of developing an implementation for a function, expressed as an algorithm that combines a collection of simpler functions.

Functional decomposition is a refinement technique that utilizes functional abstractions to manage complexity in structured, top-down approaches. Referring to Figure 4.1, this decomposition approach produces a tree-like structure in which each node f_{pqr} of the tree corresponds to a function. In this depiction of a functional decomposition, a higher level function "consists of" the functions at the next level. Although not explicit, it should be emphasized that an algorithm combines these constituent functions to produce the higher level function. In most cases, the functions have intimate knowledge of the data structures employed. As the decomposition proceeds, not only are the functions refined, but so are the data structures.

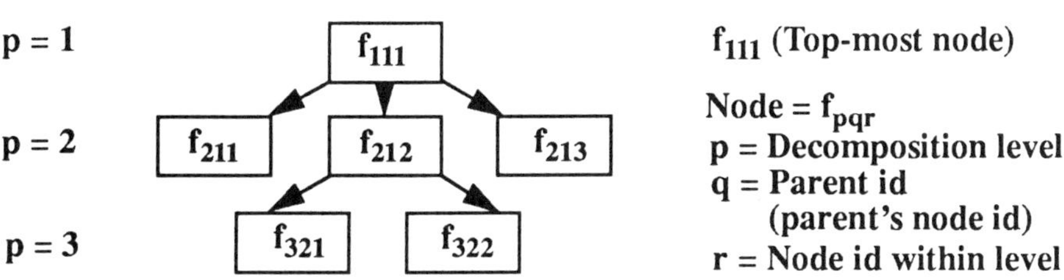

Figure 4.1 Tree-like structure produced by functional decomposition

More formally, a *decomposition D* of function f at any stage p of the process is represented by

$$D_p = \{U, V\}, \tag{4.3}$$

where U is a set of functions to be decomposed, and V corresponds to a set of leaf functions within the decomposition tree for f. In other words, the functions in V are not decomposed further. Before decomposition begins, U contains f, and V is the empty set. For the decomposition of f_{111}, this first step is depicted at the top right of Figure 4.2.

In order to perform decomposition, two operations are defined: *expansion* [112] and *selection*. In expansion, some $f_{pqr} \in U$ is replaced with an algorithm consisting of w subfunctions,

$$f_{pqr} \leftarrow f_{(p+1)r1}, f_{(p+1)r2}, \ldots, f_{(p+1)rw}. \tag{4.4}$$

More precisely, this operation corresponds to the development of an implementation, containing the functions shown in the right of equation (4.4), for the functional specification of f_{pqr}. Selection picks some functions $Z \subseteq U$ which will not be decomposed further. Thus, selection consists of the two steps shown in equation (4.5).

$$V = V \cup Z, \quad U = U - Z \tag{4.5}$$

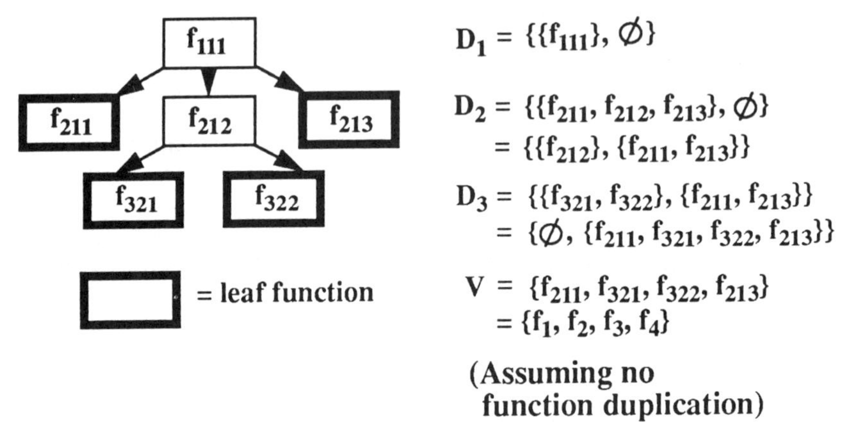

Figure 4.2 Steps in a functional decomposition

For example, in Figure 4.2, f_{111} is first replaced by f_{211}, f_{212}, and f_{213}. Next, f_{211} and f_{213} are selected as leaf functions, while f_{212} is chosen for further decomposition. Although selection was performed at this level of the decomposition, it could have been performed at the last level, after performing a series of expansions. The decomposition process terminates when U equals the empty set. At this point, V contains the most primitive (leaf) functions within the decomposition tree, represented by equation (4.6).

$$V = \{f_1, f_2, \ldots, f_y\} \tag{4.6}$$

No functions are duplicated in this equation. For simplicity, the triple subscripts have been replaced by unique indices (see Figure 4.2).

Upon performing a functional decomposition, f can be expressed as a directed graph γ consisting of nodes γ_N and edges γ_E (see equation (4.7)). As shown in equation (4.8), each node n_j corresponds to one of the leaf functions in equation (4.6). Referring to equation (4.9), an edge e_{ij} represents data and/or control flow from n_i to n_j. As will be elaborated on later in the chapter, a representation of this form can be used for hardware/software partitioning.

$$\gamma = (\gamma_N, \gamma_E) \tag{4.7}$$

$$\gamma_N = \{n_1, n_2, \ldots, n_z\}, \quad n_j \in V, \quad \forall j = 1 \ldots z \tag{4.8}$$

$$\gamma_E \subseteq \gamma_N \times \gamma_N, \quad e_{ij} = (n_i, n_j) \tag{4.9}$$

Before leaving this discussion, a final comment is necessary. Functional decomposition is one decomposition technique. Another decomposition technique based on abstract data types will be presented later in the monograph which has some advantages over functional decomposition and can also be used to derive the leaf functions for f.

4.3 VIRTUAL MACHINES

Definition 4.3 *Virtual machine*: an abstract computing element having a corresponding *virtual instruction set* that defines the operations supported by the element.

A virtual machine is an abstraction which provides a set of facilities, such as operations and resources, that can be used to define the facilities of a more abstract (higher level) virtual machine [51]. As discussed in Chapter 2, computer systems can be described in terms of layers of virtual machines. A virtual machine has the characteristics of a programmable computer, and the concept of a virtual instruction set is similar to the traditional notion of an instruction set. Therefore, a virtual machine can "execute" a program specified in terms of a sequence of operations (virtual instructions) being performed on various operands.

Note that a functional decomposition of *f* defines the functions to be provided by a virtual machine. By performing expansion and selection, the leaf functions *V* of the decomposition are derived. These leaf functions represent a virtual instruction set which defines the functions that need to be provided by a virtual machine in order to perform *f*.

Referring to Figure 4.3, data abstraction can be used to represent a virtual machine. This representation is useful for conceptualizing about virtual machines and illustrating the similarity between hardware and software elements. In this representation, the methods are equivalent to a set of virtual instructions whose execution can affect the machine's internal state. The notion of programmability is evident from this description. Specifically, the virtual instructions provided by the machine can be invoked in a particular sequence to perform some desired function.

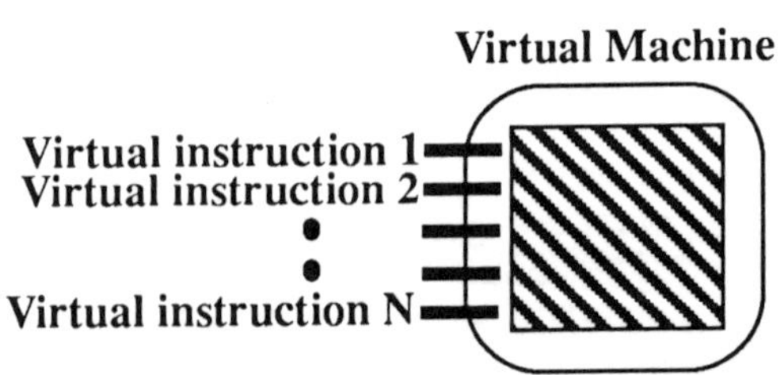

Figure 4.3 Representation of a virtual machine using data abstraction

As an example, a virtual machine called LIST may provide the virtual instruction set consisting of sort, insert, delete, and search. Insertion and deletion of elements affect the state of LIST. Another example is a virtual machine called CPU which provides the operations add, subtract, load, and store. Addition or subtraction of two elements can affect the condition codes and the internal memory, which reflect a portion of the CPU's state.

A virtual machine is a reusable entity and bears a strong conceptual relationship to a library element. Consider the construction of a software program. Suppose that a library of virtual machines was available, with each one providing a particular virtual instruction set. A program could be developed by "selecting" specific virtual instructions from various virtual machines. For example, given the virtual machines LIST and CPU above, a program might select the virtual instructions insert and sort from LIST, and the virtual instruction add from CPU. Other programs could be constructed similarly. Although program development is more complicated, this example demonstrates the usefulness of the virtual machine concept.

4.4 HARDWARE/SOFTWARE PARTITIONING

Definition 4.4 *Hardware/software partitioning*: the process of determining which functions should be implemented in (performed by) hardware and which in software.

4.4.1 Hardware versus Software Implementation

Before proceeding, a closer inspection is necessary of what is meant by implementing functions in software versus hardware. A *software implementation* consists of ISA level machine instructions and a program interpreter, a processor which fetches and executes the instructions. This definition treats the software, consisting of machine instructions, and the processor as a single unit since both elements are required to implement the function. An important characteristic of this implementation is the fetch/execute nature of the underlying processor.

In a *hardware implementation*, no ISA level machine instructions are utilized to implement the function. In this definition, microinstructions and nanoinstructions are considered part of the hardware implementation. This view is a bit controversial since microinstructions and nanoinstructions can also be treated as "software". More will be said on this point when formalizing the abstract hardware/software model. Examples of a hardware implementation include integer add performed by a functional unit within a processor and a Fast Fourier Transform performed by an ASIC.

In general, a hardware/software implementation refers to either a software implementation or a hardware implementation. This term will also correspond to a mixture of software and hardware implementations. Henceforth, where appropriate, the term implementation will encapsulate all of these ideas.

4.4.2 A Binary Decision Making Process

Suppose that a function exists for which an implementation is desired. It is possible to implement the function in software or in hardware. As illustrated in Figure 4.4, this hardware/software partitioning choice can be simplistically represented as a binary decision making process. Decisions regarding which functions should be implemented in software versus hardware may be based on performance, flexibility, or form factor (among others).

Note that hardware/software partitioning reflects a design decision but says nothing about how the function will be implemented. For example, the decision to implement a function in software does not necessarily imply anything about the software algorithm employed or the underlying processor used to execute the software. However, such a decision does impose certain constraints on the final hardware/software implementation.

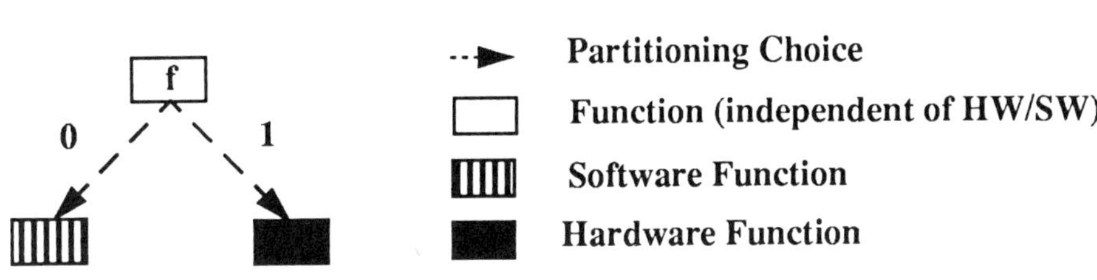

Figure 4.4 Hardware/software partitioning

4.4.3 Early versus Late Binding

An important issue in hardware/software partitioning is early versus late (deferred or delayed) binding [186]. Here, the term *binding* corresponds to a hardware/software partitioning decision. The terms *early* and *late* indicate at what stage within the design process a binding is made. In an early binding approach, hardware/software partitioning decisions are made immediately. An early binding allows the planning of the hardware/software development process to be better supported. However, once made, these decisions are difficult, if not impossible, to retract.

In general, it is difficult to make hardware/software partitioning decisions in the early stages of the design process. This difficulty in making such decisions applies particularly to new systems because so little information is available during these stages. In a late binding approach, functions are decomposed down to a suitable level before hardware/software partitioning decisions are made. Thus, in this approach, hardware/software partitioning decisions are deferred, and system functions are developed until a more intelligent determination can be made. The extent of the decomposition is dependent upon the degree of "confidence" one has in making a hardware/software partitioning decision for a given function. A late binding can help find better solutions since hardware/software trade-offs can be explored and more prudent decisions can be made with regard to the allocation of functions in hardware and in software. In addition, this approach can potentially address moving target problems, such as change requests.

4.4.4 Virtual Instruction Sets and Partitioning Granularity

The process of hardware/software partitioning is closely related to the ideas of virtual instruction sets and partitioning granularity. Figure 4.5 will be used to illustrate this relationship. In this figure, white (unfilled) boxes indicate functions which have not been allocated into hardware or software. Software functions are designated as striped boxes, and hardware functions are depicted as filled boxes. Note that the final hardware/software implementation depends on the manner in which the nodes (functions) being considered for partitioning interact, for example, in serial or in parallel fashion. In order to keep the discussion simple and focus on hardware/software partitioning, a more detailed treatment of this issue will be presented in the next chapter.

Before hardware/software partitioning (only white boxes), each of the examples in Figure 4.5 can be viewed as a functional decomposition of f down to some level (level 1, level 2, and level 3, respectively), without any commitment to hardware or software. Assuming no duplication of functions for simplicity, the leaf functions of the decomposition constitute a virtual instruction set. Thus, from left to right, the first decomposition results in a single virtual instruction, and the second produces three virtual instructions. The last decomposition corresponds to four virtual instructions. Note that the "granularity" of the virtual instructions becomes finer with additional decompositions, producing simpler functions to be considered for hardware/software partitioning.

Hardware/software partitioning can now be performed on the virtual instruction set, resulting in a binary decision making process for each virtual instruction. In the first example of Figure 4.5, the sole virtual instruction is to be implemented in hardware, which may take the form of a special purpose ASIC. In the second example, all three virtual instructions are to be implemented in hardware. In this example, if the three nodes were to interact in pipelined fashion, the final implementation would be in the form of pipelined hardware. In the last example, three virtual instructions are to be implemented in hardware while one is to be implemented in software, namely *h*.

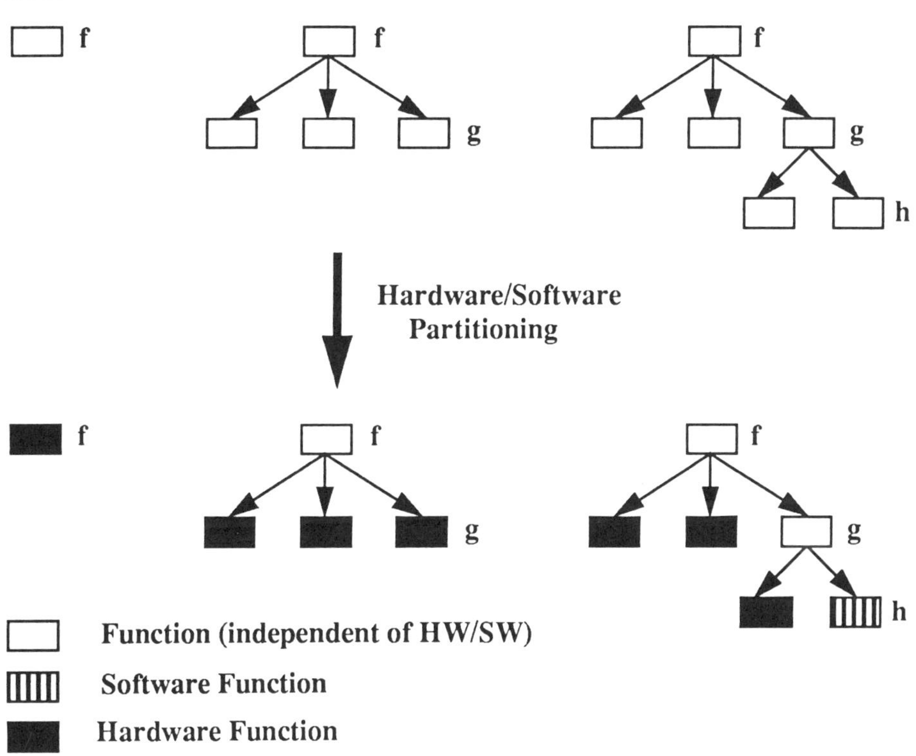

Figure 4.5 Hardware/software partitioning and level of granularity

4.5 HARDWARE/SOFTWARE PARTITIONS

Definition 4.5 *Hardware/software partition*: a description that specifies the functions to be implemented in software and in hardware, that is, a partition of the functions in γ into software functions and hardware functions.

A hardware/software partition captures the result of a hardware/software partitioning decision. In other words, it is the output of the hardware/software partitioning stage. In the discussion that follows, it is assumed that γ already exists.

In general, there can be several hardware/software partitions for some functional graph γ. A hardware/software partition $\mathcal{P}_\gamma$ for γ consists of two parts. The first part is a partition of γ_N (the nodes of γ) into *software functions* F^s and *hardware functions* F^h. Software and hardware functions correspond to functions to be implemented in software and in hardware, respectively. This partition is a functional partition of γ_N. Thus, by the definition of a set partition, the intersection of the software and hardware functions results in the empty set, and the union of the two sets produces γ_N. The next part is a description of the "interactions" between the nodes, represented by γ_E (the edges of γ). Thus, a hardware/software partition for γ can be expressed as

$$\mathcal{P}_\gamma = (\Pi_{SH}(\gamma_N), \gamma_E) = (\{F^s, F^h\}, \gamma_E) . \qquad (4.10)$$

A mathematical interpretation of a hardware/software partition can be obtained with the aid of the example in Figure 4.6. Consider a virtual instruction set consisting of three virtual instructions n_1, n_2, and n_3. Each virtual instruction can be implemented in software or in hardware. In this example, a hardware/software partition results in which two functions are to be implemented in software and one function is to be implemented in hardware.

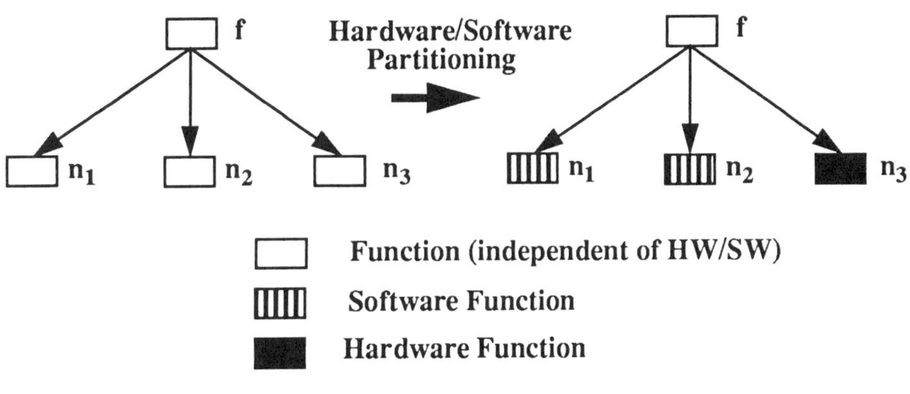

Figure 4.6 A hardware/software partition

The hardware/software partitions associated with a virtual instruction set give rise to a *lattice* [240]. A *lattice* L is a partially ordered set in which every pair of elements has a unique least upper bound (lub) and a unique greatest lower bound (glb). A partially ordered set is an algebraic structure consisting of a set X and a partial ordering relation $\leq$ on set X. The lub and glb are binary operations on elements of L as shown below.

$$lub(x, y) = x + y \tag{4.11}$$

$$glb(x, y) = x \bullet y \tag{4.12}$$

An appropriate relation can be defined such that a lattice of hardware/software partitions results. An example of such a relation is shown in equation (4.13).

$$x \leq y \leftrightarrow hardfunc(x) \subseteq hardfunc(y) \tag{4.13}$$

In equation (4.13), x and y correspond to hardware/software partitions, and the function *hardfunc(x)* extracts the set containing the hardware functions of x. This relation says that x is related to y if and only if the hardware functions within x are a subset of the hardware functions in y. In addition, the lub and glb operations can be defined as shown below. In an analogous fashion, *softfunc(x)* extracts the set containing the software functions of x. The *concat(A, B)* function concatenates the sets A and B appropriately.

$$lub(x, y) = \{concat(softfunc(x) \cap softfunc(y), hardfunc(x) \cup hardfunc(y))\} \quad (4.14)$$

$$glb(x, y) = \{concat(softfunc(x) \cup softfunc(y), hardfunc(x) \cap hardfunc(y))\} \quad (4.15)$$

A lattice of hardware/software partitions is illustrated in Figure 4.7 (only functions are shown). In the figure, the two extremes are pure hardware functions, which correspond to all functions being implemented in hardware, and pure software functions, in which all functions are implemented in software. In between, a hardware/software partition contains both software functions and hardware functions. Note the similarity between this description of a lattice and the lattice derived from the power set of a three element set Y, which represents all subsets of Y, utilizing the relation $\subseteq$ (subset).

It is important to mention that the term "pure hardware functions" reflects the partitioning decisions for the three nodes. It does not necessarily say anything about the function f. For example, f could be a software function in which the three nodes correspond to hardware functions (instructions) within a processor.

An earlier definition of a hardware/software partition included both the hardware/software partitioning decision and a corresponding design alternative expressed in terms of software and hardware [241]. The definition of a hardware/software partition provided in this section separates the hardware/software partitioning decision from the derivation of a design alternative, which is discussed in the next section.

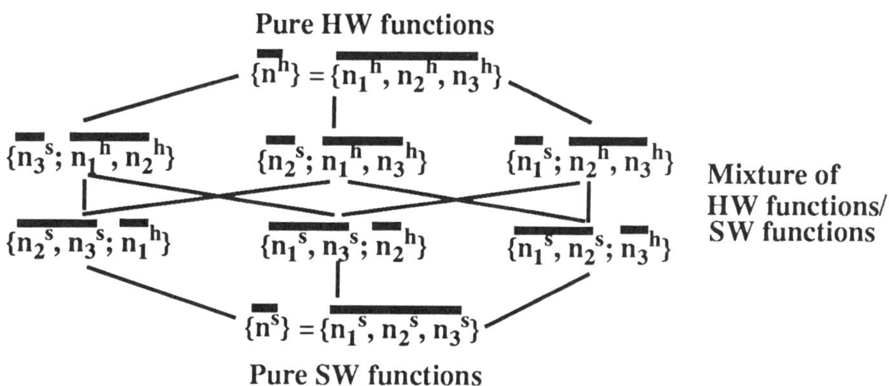

Figure 4.7 A lattice of hardware/software partitions

4.6 HARDWARE/SOFTWARE ALTERNATIVES

Definition 4.6 *Hardware/software alternative*: a possible hardware/software implementation for *f*.

A hardware/software partition is refined into a hardware/software alternative. A hardware/software alternative A for f consists of a set of *software units* S, a set of *hardware units* H, and their *communications* C. Alternatives can be described at various levels of detail and the kth alternative can be represented as

$$A_k = (S,\ H,\ C). \qquad (4.16)$$

$$S = \{\mu_1, \mu_2, \ldots, \mu_w\} \qquad (4.17)$$

$$H = \{\nu_1, \nu_2, \ldots, \nu_x\} \qquad (4.18)$$

$$C \subseteq (S \cup H) \times (S \cup H) \qquad (4.19)$$

Software and hardware units constitute the fundamental building blocks used to implement *f*. A software unit $\mu_i \in S$ represents a software/processor pair, possibly containing a scheduler. A program interpreter is an important component of a software unit. Such a unit also includes any special purpose devices, such as floating point coprocessors, which are employed as part of the interpreter's fetch/execute cycle. A hardware unit $\nu_j \in H$ corresponds to a special purpose device which is not part of a fetch/execute cycle. These devices generally accept data and produce data [242]. Three alternatives, a single hardware unit, a single software unit, or communicating software units can be represented as shown below, respectively.

$$A_1 = (\varnothing,\ H,\ \varnothing) \tag{4.20}$$

$$A_2 = (S,\ \varnothing,\ \varnothing) \tag{4.21}$$

$$A_3 = (S,\ \varnothing,\ C) \tag{4.22}$$

As seen in equation (4.23), *d* hardware/software alternatives for *f* are expressed as

$$A = \{A_1, A_2, \ldots, A_d\}. \tag{4.23}$$

A more detailed description of software units and hardware units is provided when the abstract hardware/software model is defined. Although not explicit in the notation, it should be kept in mind that a hardware/software alternative is derived from a specific hardware/software partition.

An example might help at this point to clarify these concepts. Consider the implementation of a Fast Fourier Transform function. There are several hardware/software alternatives which can be developed as possible implementations for this function. For example, one alternative is a software unit consisting of a Fast Fourier Transform program executing on a general purpose processor, such as a Motorola MC68020 processor. Another possibility is a more "specialized" software unit comprised of a program executing on a more specialized processor, such as a Texas Instruments TMS320C30 digital signal processor. As an extreme, a hardware unit, in the form of a Fast Fourier Transform ASIC, can be developed. The latter corresponds to implementing the algorithm in silicon.

As suggested above, a spectrum of alternatives is possible for the implementation of a function, and each alternative can be evaluated with respect to various metrics, such as performance, reliability, and cost. In Figure 4.8, three points, corresponding to three different hardware/software alternatives, are shown within this spectrum of possibilities. Each alternative contains a different mixture of hardware and software. The points *(X)*, *(Y)*, and *(Z)* correspond to a single software unit, a single hardware unit, and a mixture of software and hardware units, respectively.

The intent of this figure is to convey an intuitive feel for various hardware/software alternatives for a given function and provide one possible interpretation of A_k. Referring to Figure 4.8 (b), there are several points along the axis labelled *S*. As one moves along this axis to the left of point *(X)*, software units that contain more hardware functionality are encountered. For example, these software units may be more specialized. Also, some software units, which contain less software functionality (relative to *(X)*), may utilize a hardware unit to help accomplish *f*, corresponding to point *(Z)*. Note that such a software/hardware unit combination requires communication (and synchronization) of operands and results. Hardware units can also be used to implement functionality stand-alone (point *(Y)*). Finally, one can imagine constructing multiple software units or multiple hardware units which communicate and synchronize to accomplish some function.

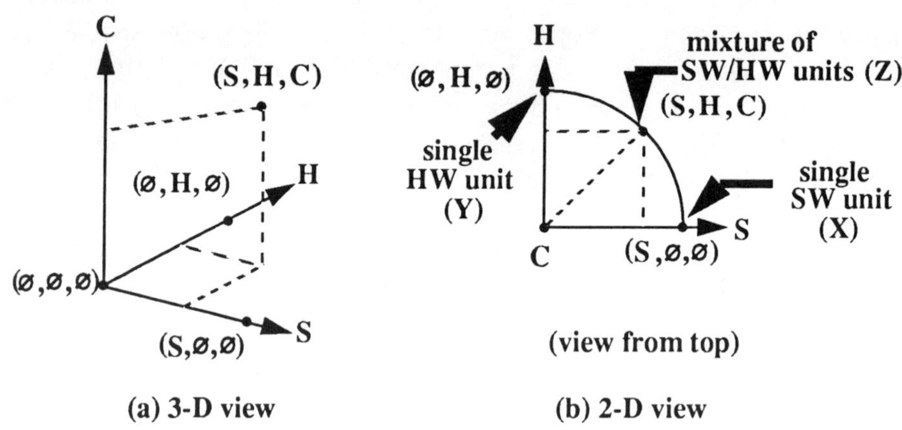

Figure 4.8 A spectrum of hardware/software alternatives

4.7 HARDWARE/SOFTWARE TRADE-OFFS

Definition 4.7 *Hardware/software trade-off* [1][6][146]: a decision regarding the allocation of functions into hardware and software that attempts to satisfy a set of criteria (objectives), many of which conflict and thus, are not attainable at the same time.

Hardware/software trade-offs are performed in an attempt to satisfy specific requirements or constraints. Some typical trade-offs include decisions regarding the implementation of a function in hardware versus software, as well as those that involve the migration of software functions into hardware (or vice-versa). These trade-offs lead to other types of trade-offs, such as performance versus cost. For example, by providing special purpose hardware within a software implementation, the overall execution time of the software may be improved. However, this increase in performance is achieved at the expense of additional hardware cost, area, power consumption, and complexity.

The concept of a virtual machine is fundamental to understanding hardware/software trade-offs. As an example, a common trade-off involves the use of floating point hardware. Consider two possible hardware/software alternatives: a "C" program compiled to execute on a Motorola MC68020 and the same "C" program compiled to execute on a MC68020 with a MC68881 floating point unit. It is worth noting that in the second alternative two virtual machines exist, one which executes "integer" instructions and another that executes floating point instructions.

Compilation of the program produces different virtual instructions for the two cases. In particular, for the program executing on the MC68020 alone, floating point operations are decomposed into the primitive operations supported by the MC68020. For the alternative which utilizes the MC68881, floating point operations are treated as primitive operations to be executed directly by the MC68881 virtual machine. At the extreme, a hardware implementation which provides the same functionality as the software implementation may be developed.

These different trade-offs are depicted in Figure 4.9. In this figure, the amount of functionality provided in hardware increases from left to right. Before proceeding, there are some important observations that need to be made. Note that the topmost node in the first two examples is a software function, as opposed to an uncommitted (white box) function independent of hardware or software (see Figure 4.6). Therefore, hardware/software partitioning has already been performed for function *f* in Figure 4.9, and the codesign problem has become one of seeking the "appropriate" mixture of hardware and software for these software implementations. Note also that a software function can be decomposed in the same manner as an uncommitted function. As a result, hardware/software partitioning can be performed on the constituent functions, thereby influencing the amount of functionality provided in hardware and in software within the implementation.

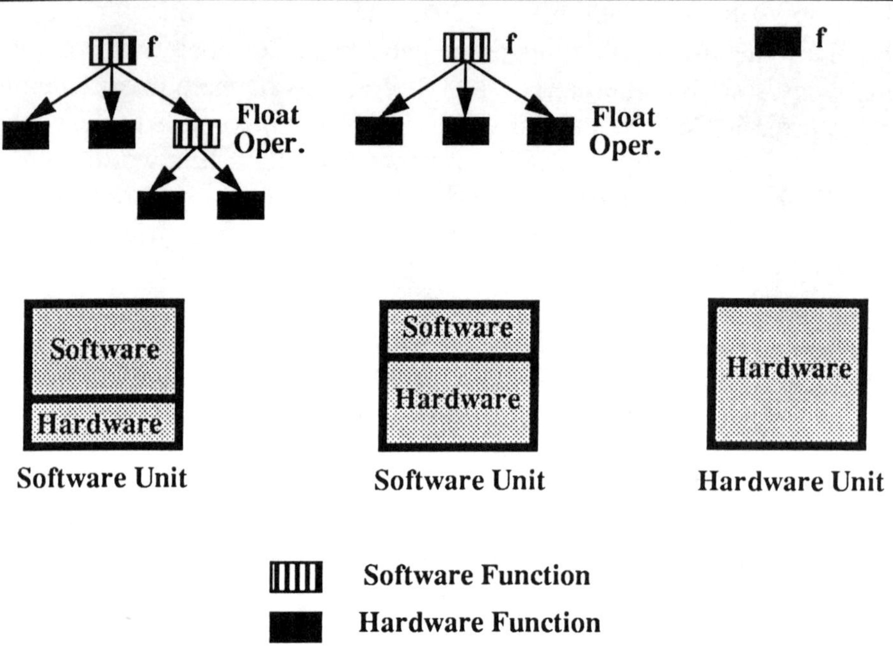

Figure 4.9 Hardware/software trade-offs

The hardware/software trade-offs illustrated in Figure 4.9 can be formally described using a collection of trade-off functions Γ as shown in equations (4.24) through (4.27). The first two equations reflect trade-offs within a software unit. These functions accept a software unit along with a software function n^s or a hardware function n^h and produce a new software unit (indicated by the prime symbol). Given a hierarchical graph description of a software program, equation (4.24) extracts the subprogram represented by n^s and embeds this functionality in hardware, while leaving the root node of n^s as a single software operation. Equation (4.25) behaves similarly except that a hardware function is expanded into a software subprogram, whose primitive operations are provided in hardware. The next two equations formalize the process of embedding the functionality associated with a software unit within a hardware unit (and vice-versa).

$$\mu_k' = \Gamma(\mu_k, n^s) \tag{4.24}$$

$$\mu_k' = \Gamma(\mu_k, n^h) \tag{4.25}$$

$$\nu_k = \Gamma(\mu_k) \tag{4.26}$$

$$\mu_k = \Gamma(\nu_k) \tag{4.27}$$

The definition of a hardware/software trade-off attempts to capture the idea that both hardware and software are affected in some way [243]. Thus, simply changing a software algorithm within an existing software unit does not constitute a hardware/software trade-off but does result in a different hardware/software alternative.

4.8 CODESIGN

Definition 4.8 *Codesign*: the cooperative development of the hardware and software portions of a system emphasizing the interaction between software and hardware design.

4.8.1 Exploring Hardware/Software Trade-offs

Design interactions between hardware and software occur in several ways. One way is through the process of performing hardware/software trade-offs. In general, numerous hardware/software trade-offs can be performed. This myriad of possibilities leads to the consideration of several design alternatives. The *exploration of hardware/software trade-offs* refers to the process of creating and evaluating hardware/software alternatives and selecting an implementation for a particular function based on some criteria.

These alternatives are evaluated with respect to various metrics. Suppose that f designates some function for which an implementation I

is required. Then, a collection of w alternatives being considered for the implementation of f can be represented by

$$I = A_1 \vee A_2 \vee \ldots \vee A_w , \quad (4.28)$$

using the definition of an alternative presented earlier. Thus, if the metric for selecting an implementation is execution time (T), an implementation I can be chosen using

$$T_I = min(T_{A_1}, T_{A_2}, \ldots, T_{A_w}), \quad (4.29)$$

or if the metric for selecting an implementation is reliability (R), an implementation I can be chosen based on

$$R_I = max(R_{A_1}, R_{A_2}, \ldots, R_{A_w}). \quad (4.30)$$

The last expression requires a unified hardware/software perspective for determining the reliability of an alternative.

4.8.2 Alternative Evaluation

Alternative evaluation requires two steps. The first step is concerned with comparing individual alternatives. The second step involves assessing the impact of the alternative on the overall system behavior. Two important issues are addressed in alternative evaluation: 1) the subtleties associated with integrating hardware/software implementations within a system and 2) the assumption that optimization of individual implementations will lead to optimization of the whole system.

To address the first step, a *quantitative evaluation model* has been developed to compare alternatives based on various metrics. Using a collection of weights, with each weight representing a number from 0 to 1 indicating the importance of a particular metric, the model determines the *quality* of an alternative, which is a figure of merit used for comparison purposes. In deriving implementations, different metrics may be important for different system functions. For example, one system function may require an implementation with minimum execution time while another may require high reliability. The weights can elevate the importance of certain metrics, affecting an alternative's quality and thus its selection as an implementation.

The quantitative evaluation model Q can be expressed using the five-tuple representation shown in equation (4.31). Given a set of alternatives A expressed in the form of equation (4.23), a set of goodness values G, and a set of weights W, the model calculates a collection of quality values κ. The set κ contains a quality value for each alternative in A. Thus, the model can be viewed as a function which associates a quality value with an alternative.

$$Q = (\rho(A), \lambda, G, W, \sigma(A, G, W)) \tag{4.31}$$

The first component of the model is a set of metric functions ρ_j (see equation (4.32)), one for each metric of interest, for example, performance, reliability, and cost. These metric functions accept the set of alternatives as a parameter. The subscript refers to a label used to denote the metric being considered. As an example, the label T may denote execution time, and C may denote cost.

$$\rho(A) = \{\rho_T(A), \rho_C(A), \ldots, \rho_X(A)\} \tag{4.32}$$

Referring to equations (4.33) through (4.37), each metric function ρ_j maps the alternatives to a quantity m_j, which represents a set of metric values, one for each alternative under consideration. Similar to metric functions, a set of goodness functions exists, denoted by λ, for each metric. Each goodness function λ_j accepts m_j and returns a quantity g_j, which corresponds to a set of values between 0 and 1 for each alternative. Each value in g_j reflects the "goodness" of an alternative with respect to a particular metric j, relative to the other alternatives. At this point, the goodness values of the alternatives are grouped together (by metric) into a set G. For example, if five alternatives were being evaluated with respect to execution time (T)

and cost (C), G would contain two elements, g_T and g_C. Each of these g_j would contain five goodness values, one for each alternative.

$$m_j = \rho_j(A) \tag{4.33}$$

$$m_j = \{mval_1, mval_2, \ldots, mval_d\} \tag{4.34}$$

$$g_j = \lambda_j(m_j) \tag{4.35}$$

$$g_j = \{gval_1, gval_2, \ldots, gval_d\} \tag{4.36}$$

$$G = \{g_T, g_C, \ldots, g_X\} \tag{4.37}$$

The weights W indicate the significance of the metrics. Each $w_j \in W$ reflects the importance of a particular metric j. The sum of the individual weights w_j is 1. Thus, for example, one may associate a weight of .4 for execution time and .6 for cost giving $w_T = .4$ and $w_C = .6$, respectively.

$$W = \{w_T, w_C, \ldots, w_X\}, \quad \sum_{\forall j} w_j = 1 \tag{4.38}$$

Finally, using A, G, and W, an evaluation function σ computes a set of quality values for the alternatives.

$$\kappa = \sigma(A, G, W) \tag{4.39}$$

The quantitative evaluation model "controls" the exploration of alternatives. Specifically, in the selection of an implementation, different metrics may be important for different system functions. The weights can elevate the significance of certain metrics, affecting an alternative's quality and selection as an implementation. Note that the weighting mechanism captures the idea of "balancing" conflicting metrics. Finally, this model can be used to evaluate alternatives at various levels of detail.

4.9 AN EXAMPLE OF ALTERNATIVE EVALUATION

To illustrate the concept of alternative evaluation, a square root function is utilized as an example. The algorithm used to implement this function is the Newton-Raphson iterative method, shown in Figure 4.10. Note that the performance of this algorithm depends on the value of the argument.

```
square_root(num: real) returns result: real
pre-condition: num >= 0
post-condition: num := result * result


square_root (num)
{
    inval = num;
    new = INIT_GUESS;
    old = DUMMY;
    diff = new - old;
    while (fabs(diff) >EPSILON) {
        old = new;
        quotient = inval/old;
        numerator = old + quotient;
        new = numerator/2.0;
        diff = new - old;
    }
    result = new;
}
```

Figure 4.10 Newton-Raphson method for performing square root

Three different instruction set level hardware/software alternatives for the square root function were developed using the ADEPT environment. The first alternative consisted of a square root program running on a general purpose processor. The second alternative consisted of a square root program running on a general purpose processor with a floating point unit. Both of these alternatives correspond to software units. The third alternative consisted of dividing the input data among two independent software units, allowing the square root of several data items to be computed in parallel.

The hardware was modeled using a mixture of library modules and custom VHDL descriptions. A model of a simple load-store machine was developed which included a probabilistic instruction cache model. A memory model contained the instructions of the square root program. The model parameters for the hardware included the processor clock cycle time, the hit ratio of the instruction cache, the read/write times from/to memory, and the delay associated with integer and floating point instructions.

Table 4.1 summarizes some information regarding the alternatives described above. Using the data inputs 100, 400, 900, 1600, 2500, and 3249, these alternatives were compared in terms of the total execution time required to perform the square root for all of the six data items. The results are shown under the column labelled T_{total} in Table 4.1. The parameters for the simulation included a clock cycle time of 100 ns, 2 clock cycles for memory read/writes, 1 cycle for instruction cache access, and a 60% instruction cache hit ratio.

The quantitative evaluation model can be used to evaluate the alternatives with respect to multiple metrics. For example, suppose that the metrics of interest are execution time (T) and cost (C). Assume that the cost of an alternative is calculated by associating a value of 1.0 for a basic software unit (software to implement functionality + general purpose processor hardware) and adding .3 for a special purpose unit. A floating point unit, employed in alternatives 2 and 3, is an example of a special purpose unit. Using this hypothetical evaluation scheme, the cost of the alternatives is shown in the column labelled C_{total} in Table 4.1.

Table 4.1 Evaluation of square root alternatives

Alternative	T_{total} (us)	C_{total}	κ^a	κ^b	κ^c
A_1	543.48	1.0	0.00	1.00	0.50
A_2	345.50	1.3	0.55	0.81	0.68
A_3	182.30	2.6	1.00	0.00	0.50

The use of a particular set of weights reflects the importance of the metrics and thus influences the quality of an alternative. For example, if only execution time was important in evaluating the alternatives above, the weight for execution time would be assigned 1, and the weight for cost would be assigned 0, resulting in the κ^a quality values in Table 4.1. On the other hand, if only cost was important, a weight of 1 would be assigned to cost and 0 to execution time, producing the κ^b quality values. If both execution time and cost were equally important, a weight of .5 would be assigned to both metrics, yielding the κ^c quality values. Thus, using the first set of weights, alternative 3 is the best choice. Using the second set of weights, alternative 1 is the best choice. Finally, alternative 2 is the best choice using the last weighting scheme.

To gain a better understanding of how the quality values are derived, this example is examined in greater detail. For illustration purposes, the set of weights utilized corresponds to case "c" (column κ^c) in Table 4.1. The set of alternatives and weights are presented below.

$$A = \{A_1, A_2, A_3\} \tag{4.40}$$

$$W = \{w_T, w_C\} = \{0.5, 0.5\} \tag{4.41}$$

As shown in equation (4.42), the set ρ consists of the two metric functions ρ_T and ρ_C for the metrics T and C, respectively. These metric functions accept the alternatives above and "generate" the values T_{total} and C_{total}, respectively, in Table 4.1. As indicated in equations (4.43) and (4.44), these values represent m_T and m_C, respectively. Using these metric values, the goodness functions, λ_T and λ_C, are used to calculate the goodness values, as shown below.

$$\rho(A) = \{\rho_T(A), \rho_C(A)\} \tag{4.42}$$

$$\lambda_T\text{: } g_T = \left(-\frac{1}{361.18}\right)m_T + 1.505, \quad \text{where } m_T = T_{total} \tag{4.43}$$

$$\lambda_C\text{: } g_C = \left(-\frac{1}{1.6}\right)m_C + 1.625, \quad \text{where } m_C = C_{total} \tag{4.44}$$

A method of deriving these goodness functions will now be described. As an example, consider the function λ_T. The metric values and the goodness values associated with execution time for the alternatives are provided in equations (4.45) and (4.46), respectively. The goodness values are derived by assigning a value of 1.0 to the alternative with the best metric value (lowest execution time), assigning a value of 0.0 to the worst alternative (greatest execution time), and plotting these goodness values against the metric values. A linear relationship is extracted, leading to equation (4.43) above. From this equation, the remaining goodness values can be obtained. A similar technique can be followed for the function λ_C (see equations (4.47) and (4.48)).

$$m_T = \{543.48, 345.50, 182.30\} \tag{4.45}$$

$$g_T = \{0.000, 0.548, 1.000\} \tag{4.46}$$

$$m_C = \{1.0, 1.3, 2.6\} \tag{4.47}$$

$$g_C = \{1.000, 0.813, 0.000\} \tag{4.48}$$

The goodness values are incorporated into a set G, which contains 2 elements as shown below.

$$G = \{g_T, g_C\} \qquad (4.49)$$

$$G = \{\{0.000, 0.548, 1.000\}, \{1.000, 0.813, 0.000\}\} \qquad (4.50)$$

The goodness values in G and the weights W can then be used by an evaluation function σ to determine the quality values for these alternatives. An example of such a function is shown below. For clarity, some additional functions are used in the description of σ. The function *extract(X,i)* returns the ith element of set X, where i refers to a sequential index of elements from left to right in the set. The element returned can be a set or a number. The function *createset(x)* creates a set consisting of the element x. Recall that $|Y|$ equals the number of elements in (cardinality of) set Y.

$$\kappa = \bigcup_{i=1}^{|A|} createset(first(i) + second(i)) \qquad (4.51)$$

$$first(i) = extract(extract(G, 1), i) \times extract(W, 1) \qquad (4.52)$$

$$second(i) = extract(extract(G, 2), i) \times extract(W, 2) \qquad (4.53)$$

For example, for $i=2$, the function *first(2)* would return the product of .548 and .5, and the function *second(2)* would return the product of .813 and .5. These two returned values would be added and a set would be created containing the single element .68 (see Table 4.1 also). Upon performing the union operation, a set of quality values results, one for each alternative. At this point, a new set of weights can be used to evaluate the alternatives.

A "C" program called *alteval* has been developed which implements the quantitative evaluation model described in this section. This program accepts two files as input: one containing a collection of metric values for a set of alternatives and another containing a collection of weights. The output of the program is a set of quality values for the alternatives.

4.10 SUMMARY

This chapter has presented several codesign concepts which will form the basis for many of the ideas presented in the monograph. The concept of a virtual machine is an important one. It provides a framework for discussions regarding hardware/software partitioning, including partitioning granularity. Data abstraction was used to represent virtual machines. This representation allows one to conceptualize about virtual machine concepts and helps to illustrate the similarities between software and hardware elements. A subsequent chapter will further clarify these ideas.

This chapter also developed the ideas of hardware/software partitions, trade-offs, and alternatives. A linear, weighted model for evaluating alternatives with respect to multiple metrics was formally described. Finally, an example of alternative evaluation using the model was presented.

Chapter 5

A Methodology for Codesign

This chapter focuses on a methodology for codesign. The opening section discusses the amount of unification present in various hardware/software design approaches. Following this discussion, the basic principles and philosophies adopted in the methodology are outlined. Next, a *framework* for hardware/software codesign [239], a codesign methodology which supports many of the concepts developed in the previous chapter, is described. The term framework implies a general "structure" whose purpose is to identify and guide important areas of investigation. Finally, an example is used to illustrate some aspects of the methodology.

5.1 Amount of Unification

A distinguishing characteristic of hardware/software design approaches is the amount of unification present between the hardware and software design processes. This unification is achieved through the use of common representations. Unified representations play an integral role in being able to support design interactions between the two domains. The degree to which unified representations are employed determines the amount of unification. Three possibilities, increasing in unification from left to right, are illustrated in Figure 5.1.

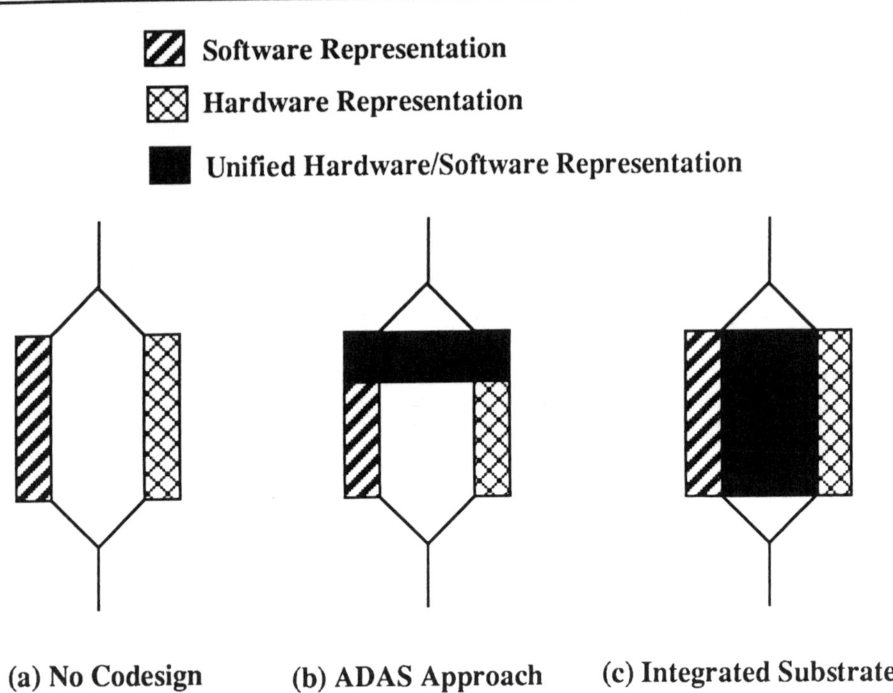

Figure 5.1 Some hardware/software design approaches

As shown in Figure 5.1 (a), one possible approach to supporting the hardware and software design processes is through the use of different tools, methodologies, and representations. Most existing hardware/software design processes utilize this approach. Although the hardware/software design process is supported, there are several limitations. One limitation is the inability to perform incremental evaluation as the design of the system proceeds. In addition, the ability to move functionality between the software and hardware domains is restricted. Finally, system evaluation occurs very late in the process. Thus, codesign is not supported in such an approach.

The ADAS codesign approach is shown in Figure 5.1 (b). In this approach, hardware/software trade-offs and evaluation are performed early in the design process using a unified representation based on directed graphs. However, from this point on, hardware and software design proceed independently until system integration, as in a typical design process.

A stronger element of unification is present in the integrated modeling substrate approach. Using this codesign approach, evaluation and trade-off exploration can be performed at several stages of the design process. Another advantage is that hardware and software designers work in familiar environments. A potential drawback is the complexity of mapping multiple paradigms onto a single, unified substrate. No known design environment has implemented this approach, and it remains to be seen whether such an approach is realizable.

5.2 General Considerations and Basic Philosophies

The intent of the codesign methodology presented in this chapter is to provide an environment for system architects, software engineers, and hardware engineers that *supports* system level hardware/software evaluation. It is desirable to allow designers the capability of exploring hardware/software trade-offs quickly and easily. Many systems are too complex to be analyzed as a whole, especially at the instruction set level. Because of this complexity, stepwise refinement approaches and the ability to incrementally add detail to a system description are necessary in the development and analysis of these systems. A first step in supporting such a methodology is to provide designers with a collection of tools, modeling constructs, and analysis techniques which assist in such an incremental refinement and evaluation.

In some circumstances, designers may have an intuitive feel for the portions of the system which deserve attention. For example, designers may have some degree of "confidence" regarding certain portions of the system, allowing these portions to be described at an abstract level. However, other portions may be new or not well understood, requiring a more detailed analysis. This discussion does not preclude the use of automated techniques, such as hardware or software synthesis. The approach is viewed as being complementary to the use of automated capabilities. In fact, some portions of the system may be synthesized immediately if so desired.

A limitation of the existing hardware/software design process is the inability to retract hardware/software partitioning decisions. Thus, an important aspect of any codesign approach is the flexibility of examining the consequences of a particular hardware/software partitioning decision (at possibly different stages of development) and, if deemed inappropriate, exploring another. One way of achieving this goal is to develop abstract hardware/software models which can be used to assess these decisions.

This idea is captured in Figure 5.2 and forms the basis for the codesign methodology presented in the next section. Although not indicated explicitly, several refinement steps may be required from the point at which hardware/software partitioning is performed to the point at which evaluation occurs. In some cases, only a portion of the system may be refined. Ideally, the dashed path would not be necessary.

There are several benefits of this approach. One benefit is the ability to evaluate hardware/software systems quickly using abstract models, as opposed to detailed, instruction set level models. This evaluation is possible due to the use of a unified hardware/software model which can be refined within a common simulation environment. As a result, common analysis techniques can be utilized to examine such aspects as reliability and performance. For example, bottleneck analysis can reveal software functions which require improvement or perhaps hardware support. Also, this approach does not require that all descriptions be provided at the same level of detail (such as the instruction set level) before any evaluation is allowed. The use of an integrated environment allows the consequences of different hardware/software decisions to be evaluated within the context of the system being designed and supports model continuity, the gradual migration of system models into hardware/software implementations.

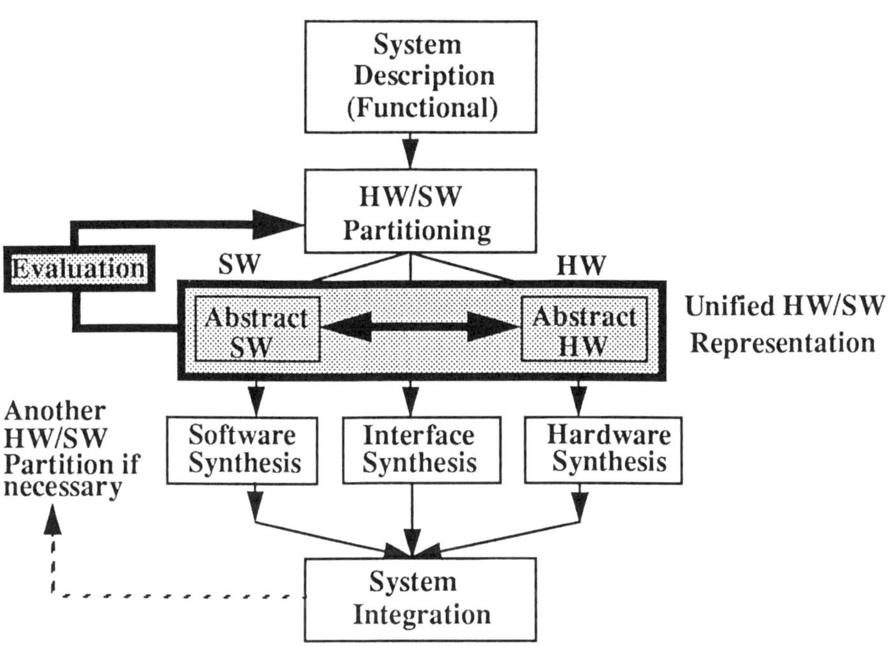

Figure 5.2 A methodology for early hardware/software evaluation

5.3 A FRAMEWORK FOR CODESIGN

Figure 5.3 depicts a general system design methodology [239] supporting the concepts presented in the previous chapter. The methodology attempts to capture aspects of both the ADAS and integrated substrate approaches to codesign. It is iterative in nature and serves to guide codesign exploration. Although not explicit in some cases, it is possible to go back to a previous step in the methodology, for example, from codesign to system partitioning. Note that a hardware/software alternative can be derived in several ways. One approach is to perform hardware/software partitioning on a functional description γ and then refine the resulting partition $\mathcal{P}_\gamma$ into an alternative A. Another possibility is to refine an existing hardware/software partition in a different manner, producing a new alternative for consideration.

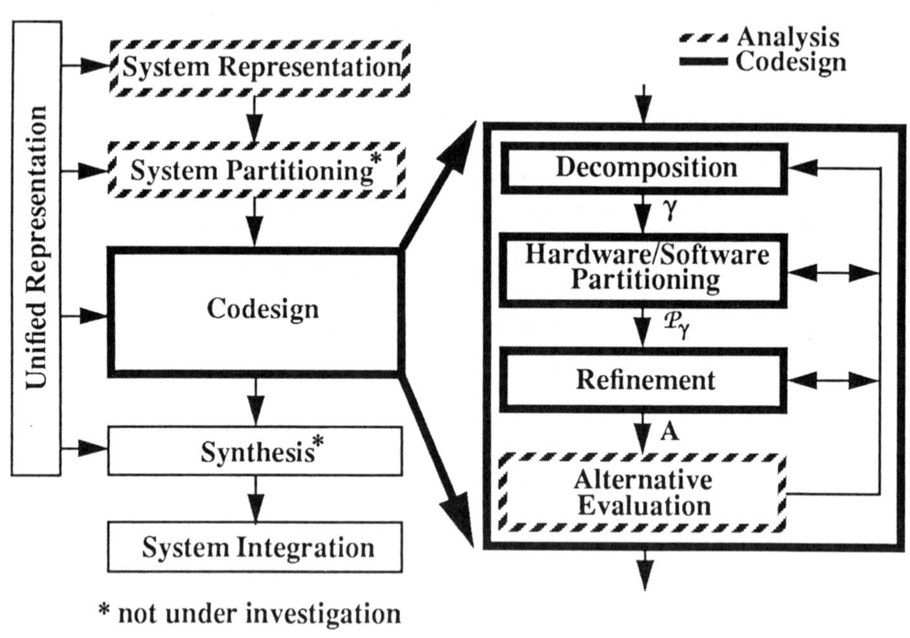

Figure 5.3 System design methodology supporting codesign

ADEPT can be used to provide one possible implementation of the methodology. The ADEPT modules serve as a unified representation for several stages of the design process. A more detailed discussion of the methodology is now presented.

5.3.1 System Representation

As shown in equations (5.1) through (5.3), a system Ψ is described as a directed graph consisting of one or more *system functions* to be performed, independent of hardware or software. It is implicit that decomposition techniques are applied at this stage. The system functions interact via some mixture of data/control flow. The superscript F in equation (5.1) emphasizes the functional nature of the description at this point.

$$\Psi^F = (N,\ E) \tag{5.1}$$

$$N = \{f_1, f_2, \ldots, f_v\} \tag{5.2}$$

$$E \subseteq N \times N,\quad e_{ij} = (f_i,\ f_j) \tag{5.3}$$

It is important that the manner in which the system functions interact, for example, serial, parallel, or pipelined, be captured since the resulting system description influences later steps of the methodology, such as hardware/software partitioning. This ability to maintain consistent descriptions of the system at various stages of development aids in supporting model continuity. As an example, consider a system level description consisting of two functions interconnected via a finite length buffer. This system representation implies that the functions execute concurrently during certain periods of time. Thus, if both of the functions were implemented in software, two processors would be required, one for each software function. However, if the functions were to interact in a serial fashion (with no buffer), it would be possible to share a single processor.

A system description can be represented in different ways. Referring to Figure 5.4, some system level models are described with only data flow, in which data flows into and out of each functional element f. The incoming flow represents operands to be processed by the function, and the outgoing flow corresponds to results generated by the

function. Functional elements execute when data is present on the inputs. In some circumstances, a mixture of data and control flow may be used to model a system. Regardless of the representation employed, it should be observed that the nodes can be viewed as a collection of virtual instructions.

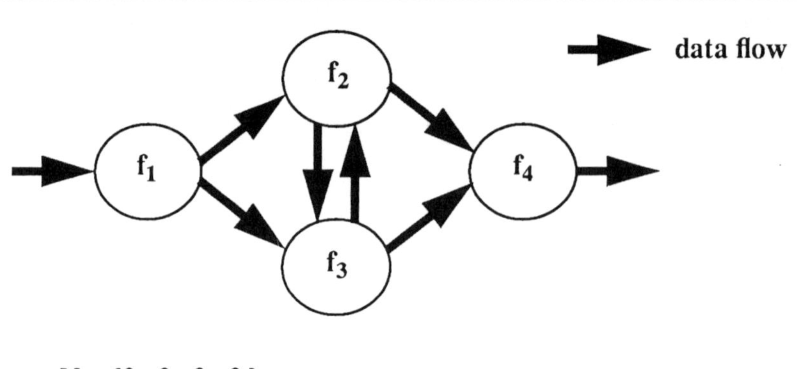

$N = \{f_1, f_2, f_3, f_4\}$
$E = \{(f_1, f_2), (f_2, f_3), (f_3, f_2), (f_1, f_3), (f_2, f_4), (f_3, f_4)\}$

Figure 5.4 A system representation containing data flow

Performance and reliability analysis can be utilized to help establish constraints for later stages. For example, system constraints may be expressed in terms of throughput, execution time, and reliability. Also, hardware constraints, such as area and power consumption, can be used to help guide later design decisions. Software constraints may be expressed in terms of various metrics, such as complexity or number of lines of code. In addition, functional specifications would be employed to refine the system functions into more detailed descriptions. These constraints (budgets) and specifications can be captured in a *specification template* for each system function and for the system as a whole. The specification template can be viewed as an "attribute" of a node in the system representation.

5.3.2 System Partitioning

In this step of the methodology, the system functions are mapped onto separate physical units, such as chips or boards. For example, in Figure 5.5, system functions f_2 and f_3 may be performed on one board while system functions f_1 and f_4 may be performed on another board. Thus, referring to equations (5.4) through (5.6), system partitioning results in the creation of mutually disjoint *system blocks* B_j. One criteria used to perform this partitioning may be the minimization of communication between system functions based on techniques found in [168].

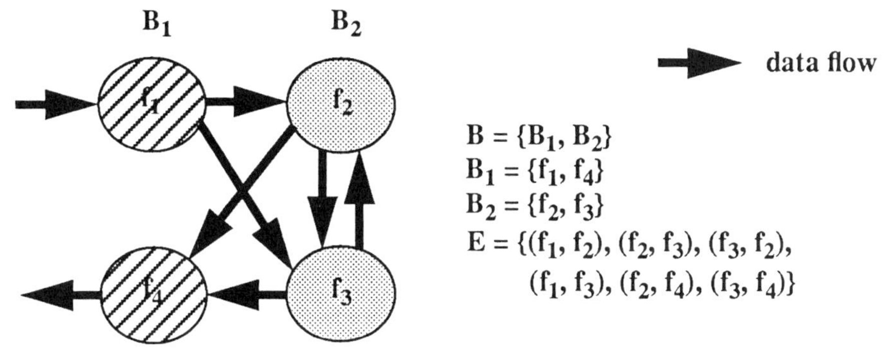

Figure 5.5 System partitioning

$$\Psi^B = (B, E) \tag{5.4}$$

$$B = \{B_1, B_2, \ldots, B_t\}, \quad B_j \subseteq N, \quad \bigcup_{j=1}^{t} B_j = N, \quad \bigcap_{j=1}^{t} B_j = \varnothing \tag{5.5}$$

$$E \subseteq N \times N, \quad e_{ij} = (f_i, f_j) \tag{5.6}$$

It should be emphasized that this step of the methodology corresponds to a *functional* partitioning of the system. In some descriptions of hardware/software design methodology, system partitioning is used synonymously with hardware/software partitioning. In this monograph, system partitioning and hardware/software partitioning refer to different stages of the design process. Note that system partitioning can also influence later steps of the methodology. For example, it is unlikely that software functions in two different system blocks would share hardware.

5.3.3 Codesign

Codesign consists of iteratively performing decomposition, hardware/software partitioning, refinement, and hardware/software alternative evaluation for each system function. At the conclusion of this process, an implementation I is derived for the system, consisting of implementations for each of the system functions as shown below.

$$\Psi^I = (I,\ E) \tag{5.7}$$

$$I = \{I_1, I_2, \ldots, I_v\} \tag{5.8}$$

$$E \subseteq N \times N,\quad e_{ij} = (f_i,\ f_j) \tag{5.9}$$

Decomposition entails deriving a virtual instruction set for each system function and expressing the function as a graph γ. Note that γ can also be the graph which describes the information flow between the system functions. In this situation, no further decomposition of the system functions is necessary, and hardware/software partitioning can be performed directly on this graph.

Hardware/software partitioning determines which of the virtual instructions in γ should be implemented in hardware and which in software. Refinement creates an abstract hardware/software model, representing a design alternative at some level of detail. Software functions may be decomposed and mapped onto one or more abstract processors [241]. The model employs a unified representation for hardware and software. Alternative evaluation uses the quantitative evaluation model to assess the goodness of individual alternatives. This evaluation process can also be used for risk assessment, as in the spiral model of software development [115][116].

The abstract hardware/software model can be utilized in several ways. The effects of various changes to an alternative can be evaluated, such as the incorporation of faster hardware resources. Also, the evaluation of different algorithm implementations of a software function on a given processor (and vice-versa) is possible.

As shown in Figure 5.6, the abstract hardware/software model can potentially be used to support the concept of an integrated substrate during the development of hardware and software. As hardware/software development proceeds (towards the right in Figure 5.6), the hardware/software model (**HSM**) can be refined in an integrated fashion. In addition, other types of hardware and software descriptions can be "mapped onto" this unified model. Using this approach, evaluation and trade-off exploration can be performed at multiple stages of the design process.

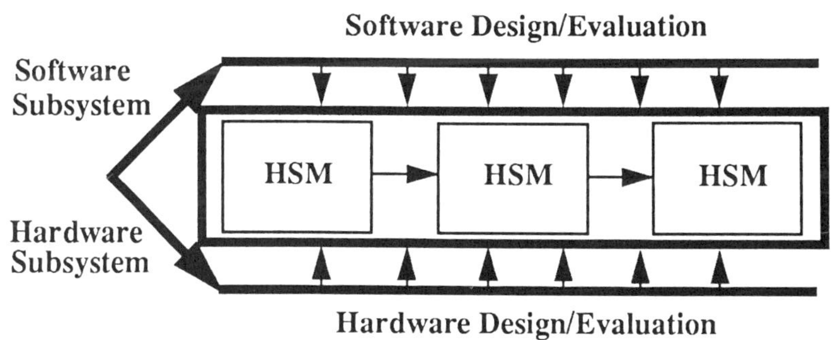

Figure 5.6 The hardware/software model as an integrated substrate

5.3.4 Synthesis and System Integration

During synthesis, hardware realizations are generated from the descriptions obtained in the codesign step. High-level synthesis, logic synthesis, and module generation tools would be utilized for this purpose. Software synthesis techniques are also employed to generate the system software. Finally, system integration requires running the synthesized software with the synthesized hardware. At this point, the system is evaluated as a complete unit.

5.4 Methodology Discussion

The codesign methodology proposed in this section is quite general. Both early and late binding are supported. Also, the methodology allows one to explore various hardware/software partitioning decisions and retract these decisions if found to be inappropriate. Note that the hardware/software partitioning step can be performed recursively, allowing partitioning to be performed on system functions or software functions.

As suggested by Figure 5.6, a unified hardware/software model can potentially rectify inappropriate hardware/software partitioning decisions at a *late* stage in the design process. The hardware/software model provides an additional level of flexibility by allowing functions to be moved between the software and hardware domains at several stages of the development process. This capability is perhaps most applicable to situations in which poor decisions were made using an early binding approach. Of course, the ability to perform hardware/software trade-offs late in the design process assumes a flexible approach to hardware and software design.

The codesign portion of the methodology incorporates the essential ingredients that influence the derivation of a hardware/software alternative. The decomposition step plays an important role in determining the mixture of hardware and software within an alternative [244]. Decomposition manages complexity and defers decisions regarding the selection of functionality in software versus hardware. The decomposition of a function can be performed down to an arbitrary level, which affects the hardware/software partitioning granularity. A large amount of decomposition, that is, several levels, results in simpler virtual instructions. A small amount of decomposition produces more complex virtual instructions. Thus, a virtual instruction can correspond to a complicated function, such as a Fast Fourier Transform, or a simple function, such as a multiply.

The virtual machine approach to codesign is general. This approach can be used to derive application-specific implementations that consist of general purpose processors, specialized processors, special purpose hardware, and their software. The virtual instructions associated with a decomposition of a function can represent operations within a software program or operations to be performed by a hardware element. For example, the decomposition can be treated as a software program which is to be executed on either a general purpose processor or a more specialized processor (two different kinds of software units). Alternatively, the decomposition can be converted into special purpose hardware (hardware unit) whose resources provide the functionality specified by the virtual instructions. Thus, the methodology supports the design of heterogeneous systems using a homogeneous framework.

5.5 An Example

To further illustrate the ideas discussed in this chapter, an example is presented. Suppose that an application has been decomposed into the system functions shown in Figure 5.7. The *source* and *sink* are used to model the production and consumption of data, respectively. For example, the source may represent a sensor, and the sink may correspond to another subsystem which further manipulates the data. It is assumed that the system functions are "buffered", and thus, the functions exhibit pipelined behavior.

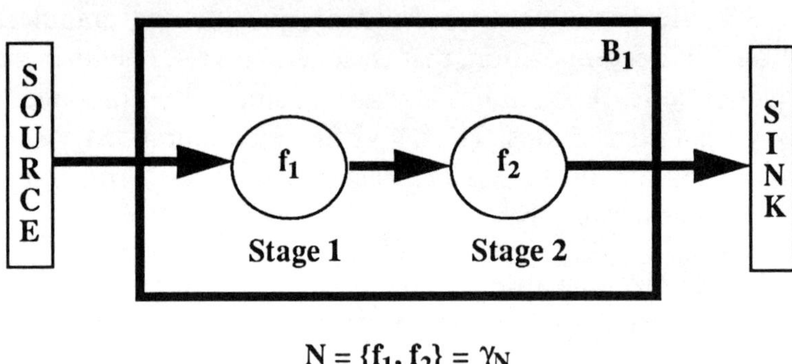

$N = \{f_1, f_2\} = \gamma_N$
$E = \{(f_1, f_2)\} = \gamma_E$

Figure 5.7 An example system

The system can be described in terms of equations (5.10) through (5.12). It should be emphasized that this view of the system is independent of hardware or software.

$$\Psi^F = (N, E) \tag{5.10}$$

$$N = \{f_1, f_2\} \tag{5.11}$$

$$E = \{(f_1, f_2)\} \tag{5.12}$$

For simplicity, all system functions are mapped onto a single system block B_1, such as a board.

$$\Psi^B = (B, E) \tag{5.13}$$

$$B = \{B_1\} \tag{5.14}$$

$$B_1 = \{f_1, f_2\} \tag{5.15}$$

$$E = \{(f_1, f_2)\} \tag{5.16}$$

If interested in performance, one could analyze this system level model in terms of throughput by associating various delays with the system functions. The analysis can be used to establish constraints for the system functions. Upon analyzing the system throughput, one can explore various hardware/software partitions. For this example, the system level model will be considered γ, which consists of two functions to be considered for hardware/software partitioning. One may decide to implement both functions in software, resulting in the hardware/software partition shown below.

$$\mathcal{P}_\gamma = (\{F^s, F^h\}, \gamma_E) \tag{5.17}$$

$$F^s = \{f_1, f_2\} \tag{5.18}$$

$$F^h = \varnothing \tag{5.19}$$

The two software functions can be refined into hardware/software alternatives. During this process, the appropriate mixture of hardware and software required to satisfy specific objectives would be explored. The alternatives would be evaluated using the quantitative evaluation model and an implementation selected for each system function. Because of the concurrency present in the model, two processors will be required in this implementation. Thus, as indicated in Figure 5.8, the final hardware/software implementation would consist of two communicating software units. As suggested by the figure, the amount of software and hardware used to implement the system functions may be different. In this example, each software function is mapped onto a single processor.

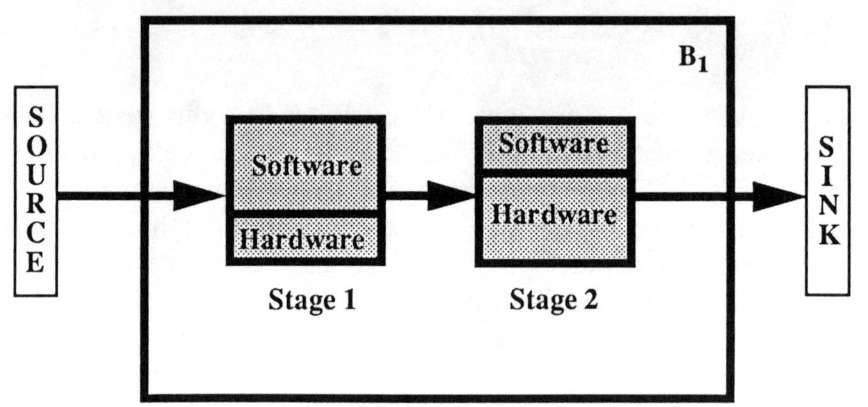

Figure 5.8 One implementation for the example system

Another possibility one may wish to explore is the implementation of f_2 as a hardware function. This hardware/software partition corresponds to equations (5.20) through (5.22).

$$\mathcal{P}_\gamma = (\{F^s, F^h\}, \gamma_E) \quad (5.20)$$

$$F^s = \{f_1\} \quad (5.21)$$

$$F^h = \{f_2\} \quad (5.22)$$

Referring to Figure 5.9, an implementation consisting of a single software unit and a single hardware unit would be produced.

Note that the view of an alternative is recursive. Specifically, one can consider alternatives for the entire system. Also, one can consider alternatives for a particular system function.

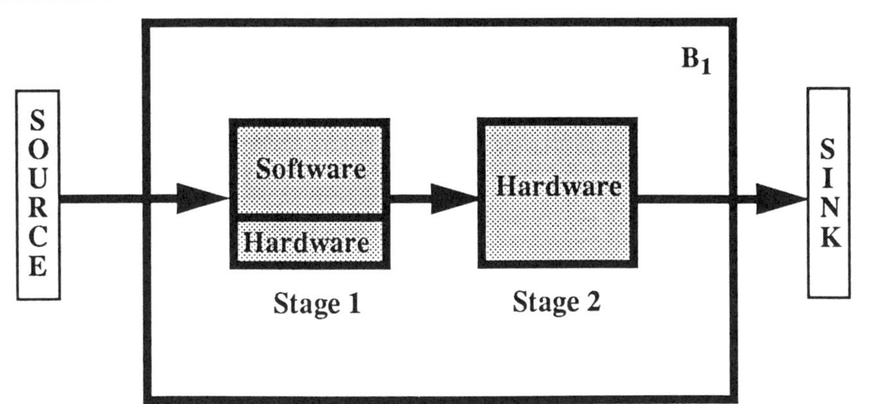

Figure 5.9 Another implementation for the example system

5.6 SUMMARY

This chapter has presented a codesign methodology that supports early performance evaluation and trade-off exploration. At the core of this methodology lies a hardware/software model employing a unified representation. This model can be used in two ways. An abstract hardware/software model can be employed for evaluating hardware/software partitioning decisions before committing to a particular design. Another possibility is to use the model as an integrated substrate such that evaluation and trade-off exploration can be performed incrementally. Thus, the methodology incorporates ideas from both the ADAS and integrated modeling substrate approaches.

An interesting aspect of the methodology is the ability to use alternative evaluation as a means of performing risk assessment, an idea borrowed from the spiral model of development utilized in software engineering. This approach allows an engineer to evaluate the consequences of design decisions using an abstract hardware/software model described at an appropriate level of detail.

Chapter 6

A Unified Representation for Hardware and Software

From this point forward, the discussion will focus on hardware/software modeling. As a point of departure, this chapter presents a unified representation [245] for hardware and software. The first section reviews the benefits of a unified representation. The next section introduces some pertinent modeling concepts, concentrating on two in particular: level of abstraction and level of interpretation. Both abstraction level and interpretation level affect the amount of detail present within a model. A unified representation is developed which incorporates descriptions based on either functional abstractions or data abstractions and integrates the modeling concepts mentioned above. This representation serves as the foundation for models that utilize data/control flow concepts (Chapter 7/Chapter 8) and object-oriented techniques (Chapter 9).

6.1 Benefits of a Unified Representation

Figure 6.1, copied from Chapter 1, summarizes many of the benefits associated with a unified representation. All of these benefits are due to the similarity between hardware and software. A unified representation addresses the current separation between the hardware and software design processes. Common design and analysis techniques, such as those used to determine performance, reliability, complexity, and correctness, can be employed. In the same vein, such a representation offers the potential of utilizing a uniform design methodology for both hardware and software, one which makes use of the techniques listed above. A unified representation also allows the possibility of cross fertilization between the software and hardware domains. Therefore, techniques and results from one domain can be applied to the other. Another benefit is that synthesis can be performed from a common design representation. This idea has been exploited in cosynthesis capabilities.

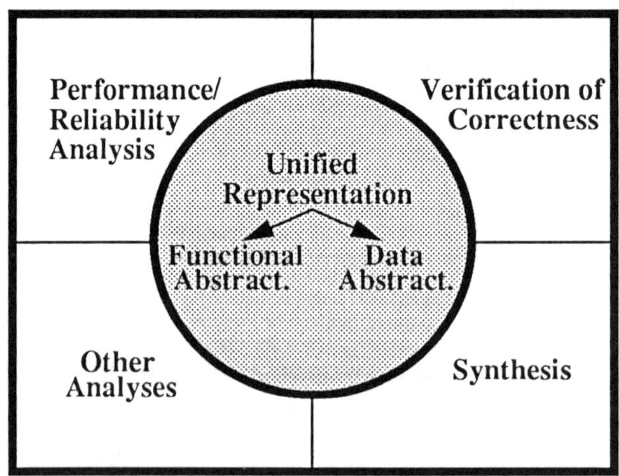

Figure 6.1 Benefits of a unified representation

Several other advantages exist as well. A unified representation supports the early evaluation of hardware/software systems in a common simulation environment and thus, enhances the communication between hardware and software developers through the use of a common modeling paradigm. Also, hardware/software trade-offs can be performed more easily by allowing functionality to be transferred between software and hardware.

6.2 MODELING CONCEPTS

Before presenting the unified representation, it is necessary to introduce some important modeling concepts. These concepts include definitions of a model, abstraction, interpretation, and a discussion of the difference between abstraction and interpretation.

6.2.1 Models

Definition 6.1 *Model* [246][247]: a description of an entity, provided at some level of detail, that conveys the important properties of interest.

As illustrated in Figure 6.2, the term *entity* (called *prototype* in [246]) refers to a specific aspect of reality and corresponds to an object, a situation, or a system. A model *M* is associated with a particular entity. Models play an important role in learning about entities in the world around us. In many circumstances, the entity may be too big, too complicated, or perhaps even too dangerous to examine. Models aid in understanding and analyzing the entity's behavior without having to deal with the intricacies of the entity. For example, engineers construct physical models, such as wind tunnels, to understand various properties of aerodynamics, or mathematical models, expressed as a collection of equations, which aid in characterizing such things as the movement of physical objects.

Models are important for managing complexity. In fact, an important rationale for the development of models is to accommodate our human limitations in dealing with complexity [247]. Thus, models are developed which focus on certain macroscopic properties and ignore inessential details. For example, in hardware, a processor model may only describe how the internal state is accessed and manipulated. At this level of detail, other aspects, such as the organization of the registers and the functional units, can be ignored. In software, abstract models may be constructed which utilize concepts such as matrix, list, and table, without worrying about how these concepts are to be implemented. In some cases, a given level of detail may not be adequate to examine the attributes of interest and further detail in the model may be required.

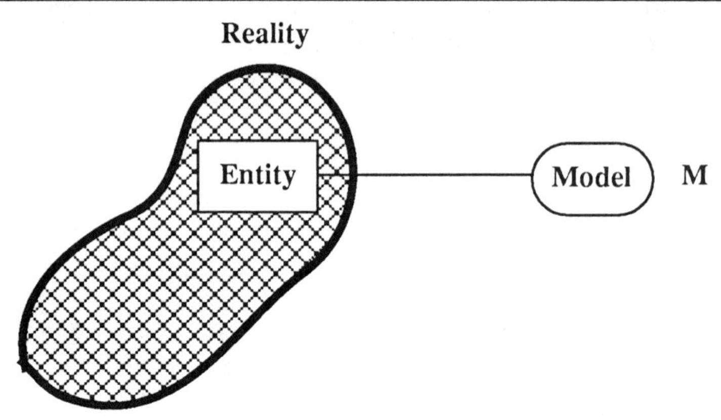

Figure 6.2 Relationship between entity and model (From [246], © 1984. Reprinted by permission of John Wiley & Sons, Ltd.)

In general, models are an approximation of an entity. As a result, a model may not always characterize all of the properties of the entity. In many cases, certain assumptions are made in the development of a model, and the validity of these assumptions directly influences how well the model characterizes the entity. However, it is important to judge a model, not necessarily in terms of right or wrong, but in terms of whether the model is adequate to reveal certain insights about the entity [246].

6.2.2 Abstraction

Definition 6.2 *Abstraction* [117]: 1) the process of ignoring (hiding) inessential details and focusing on (exposing) relevant properties of interest; 2) a description arrived at through such a process.

Definition 6.3 *Level of abstraction*: the amount of implementation information that is both hidden and exposed within a description.

As suggested by the discussion in the previous section, the concept of a model is closely related to the definitions regarding abstraction. Specifically, a model is an abstraction of an entity, and thus, abstraction is used in the construction of a model. Models exist at different levels of abstraction l^a. Therefore, a model M is a function of l^a, expressed as $M(l^a)$, and moving to lower levels of abstraction increases the detail within the model. Functional or data abstractions can be used in the development of a model.

Two methods of abstraction are *abstraction by parameterization* and *abstraction by specification* [117]. These methods apply to both functional and data abstractions. Abstraction by parameterization allows a set of computations to be represented. The term computations is being used here in a general sense. For example, in software, a search function may be parameterized with the array and element to be searched, allowing several different searches to be described. In hardware, a model of an arithmetic logic unit (ALU) can be parameterized with input operands and the operation to be performed, allowing several operations on various operands to be represented.

Recall that a specification describes the behavior of an abstraction, and an implementation realizes the behavior. Abstraction by specification allows the implementation to be ignored and promotes information hiding [131]. In the case of the search function above, only the input-output behavior may be of interest, not the algorithm used to perform the function. In the same way (see Figure 6.3), the implementation of an ALU in terms of lower level primitives, for example, AND gates, OR gates, and inverters, can be abstracted away.

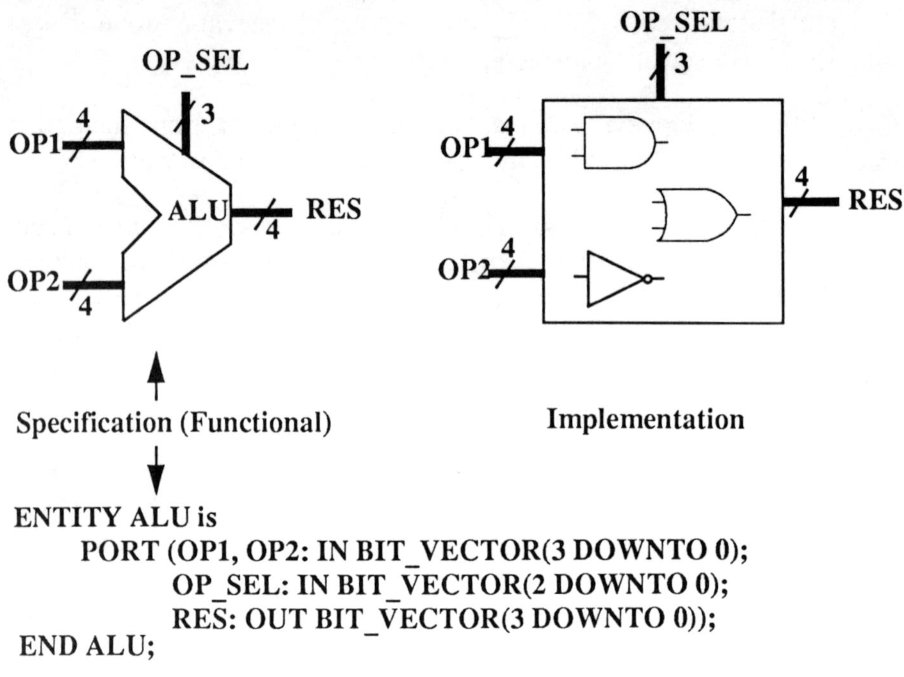

Figure 6.3 Abstraction by specification

6.2.3 Interpretation

Definition 6.4 *Interpretation* [90][91]: the process of associating semantics, particularly values and functional information, to various elements of a model.

Definition 6.5 *Level of interpretation*: the extent to which semantics are associated with a model's elements.

A continuum of interpretation levels l^i is possible within a model. The extreme levels are *uninterpreted* and *interpreted.* Uninterpreted models, such as Petri nets and queuing models, are used for system level performance and reliability analysis. These models describe the information flow within a system and, in their purest form, lack function. Uninterpreted models used for performance analysis contain temporal information in the form of probabilistic or fixed delays.

At the other extreme, interpreted models are used for detailed design and include functional information. In the most general case, these models contain both functional as well as temporal information. Thus, additional detail exists in interpreted models relative to uninterpreted models. Roughly speaking, the functional correctness of an uninterpreted model cannot be verified whereas it can for an interpreted model. Finally, a *hybrid model* [21][92][248] consists of both uninterpreted and interpreted models within a common simulation environment.

Uninterpreted and interpreted models were explored by Agerwala [249] and Auletta [92]. In his study of parallel systems, Agerwala divided models into three categories: uninterpreted, semi-interpreted, and interpreted. Based on this work, Auletta formally defined an uninterpreted model which could be implemented in a hardware description language, such as VHDL. In addition, Auletta refined Agerwala's classification of models into uninterpreted and interpreted, based on whether the model lacked or included functional transformations, respectively.

A slightly modified, but equivalent, description of the uninterpreted and interpreted model definitions of Auletta is now presented [92]. Given a model consisting of a finite set of elements $M = \{e_1, e_2, \ldots, e_n\}$, the model is uninterpreted if the following conditions are true:

(a) There exists one or more controlling mechanisms Cm within an element which indicate whether the element is active.

(*b*) A set of storage variables for the model $S = \{s_1, s_2, \ldots, s_k\}$ is defined.

(*c*) The initial values of the variables in S are defined.

(*d*) For each e_i, an input/output set is specified which is a subset of S.

If the previous conditions are true and the following condition holds, the model is considered interpreted.

(*e*) There exists a set of functions $F = \{f_1, f_2, \ldots, f_n\}$, one for each element e_i, which transforms the values in the input set to the output set.

For an uninterpreted model, the controlling mechanism *Cm* is used to determine if an element is active, in which case a token is placed on the element's output. The initial values of the storage variables can be defined. However, because of the lack of functional transformations (condition (e)), the simulation of an uninterpreted model may encounter undecidable conditions. Specifically, undecidable conditions will arise when the model is required to perform some action based on the value of a variable. Thus, some "rule" must exist for resolving the action to be taken. A simple example of such a situation is the execution of a branch instruction which tests, for example, whether the value of a register equals zero. This problem is referred to as the *test on value problem* [92].

To better understand the difference between uninterpreted and interpreted models, consider the ALU in Figure 6.4. Figure 6.4 (a) corresponds to an uninterpreted model of the ALU in which tokens arrive at all of the inputs, and the output token is produced after a fixed delay, corresponding to an ADD computation. The *JUNCTION* module fires only when tokens are present on all of its inputs. After firing, the token is delayed by the *FIXED_DELAY* module for 5 nanoseconds (ns) before being output.

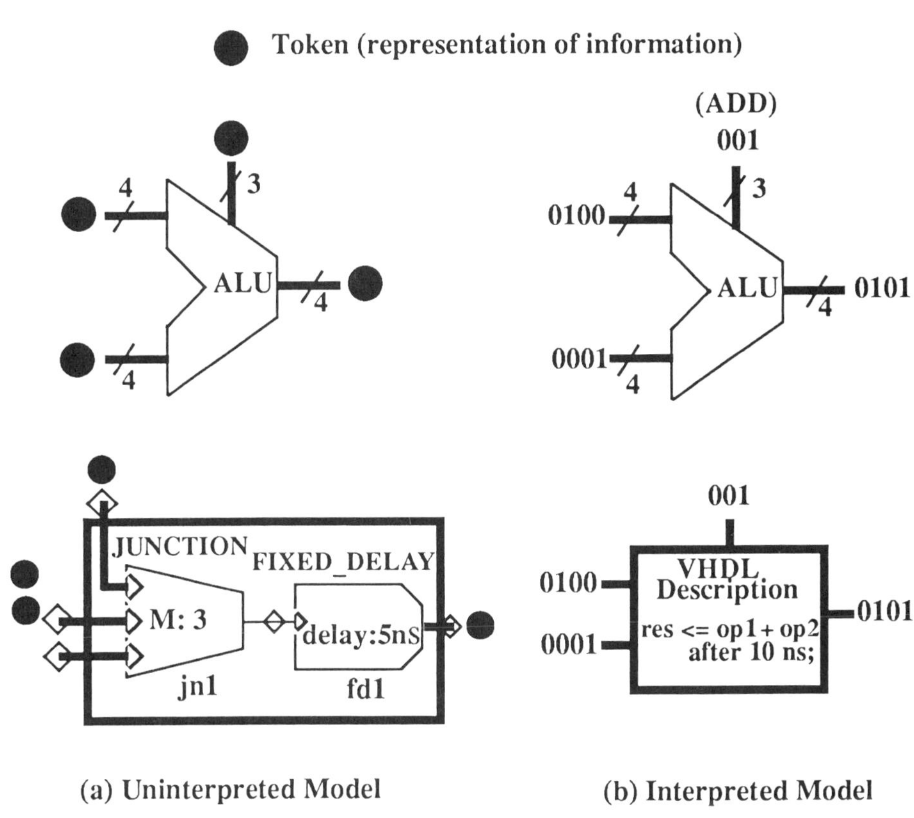

Figure 6.4 Uninterpreted versus interpreted models

A portion of an interpreted model of the ALU is shown in Figure 6.4 (b). The model is a VHDL behavioral description which contains both functional transformations that operate on actual values as well as temporal information. In some circumstances, the values are simply inspected for decision-making purposes, as in the case of the input which specifies the ADD operation ("001"). In other circumstances, functional transformations are performed on input values, such as the input operands "0100" and "0001". Note that values can be considered "colored" tokens.

6.2.4 Abstraction versus Interpretation

As indicated in Figure 6.5, level of abstraction and level of interpretation are orthogonal concepts. In particular, at a given level of abstraction, an uninterpreted or an interpreted model can be developed. As illustrated in Figure 6.4, it is possible to have an uninterpreted or an interpreted model of an ALU described at the register-transfer level of abstraction. Both level of abstraction and level of interpretation influence the amount of detail present in a model. As a result, a model can be expressed in terms of both l^a and l^i, that is, $M(l^a, l^i)$.

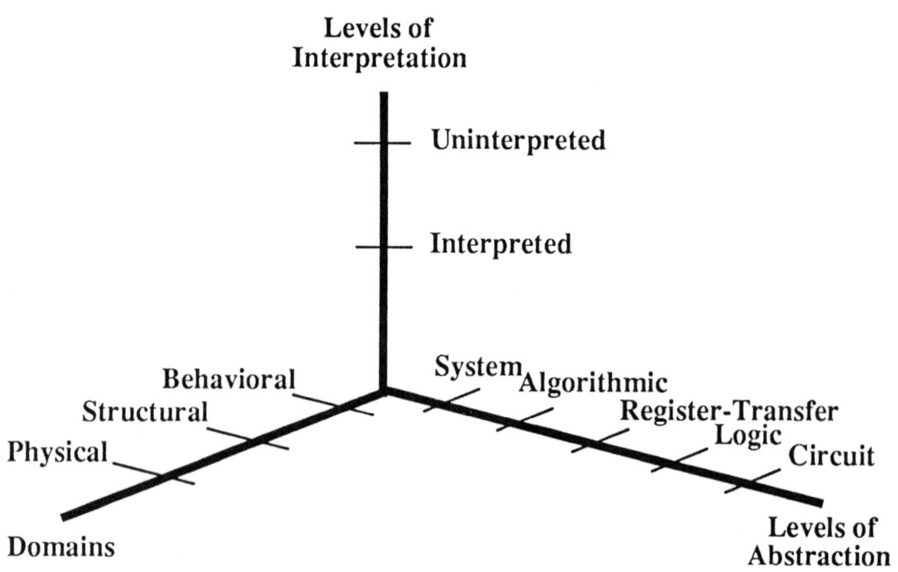

Figure 6.5 Abstraction versus interpretation

6.3 A UNIFIED REPRESENTATION

Definition 6.6 *Unified representation:* a representation that can be used to describe hardware or software.

Functional abstractions or data abstractions can be utilized in the development of a unified representation. As an example of functional abstractions, data/control flow graphs, which contain nodes that represent operations and arcs that correspond to data and/or control flow between nodes, provide an abstract representation of an element. Each node corresponds to a functional specification whose abstract implementation can be expressed in terms of another, more detailed data/control flow graph.

As illustrated in Figure 6.6, data abstractions, commonly employed in software development, can also serve as a unified representation for both hardware and software [250]. It is natural to think of hardware resources as components which consist of state and operations that manipulate this state. Thus, hardware components can be modeled using abstract data types in the same manner as software elements. In addition, object-oriented concepts, such as the notion of component classes and specialization through inheritance, apply readily to the hardware domain. A more detailed discussion of this unified perspective is provided in Chapter 9.

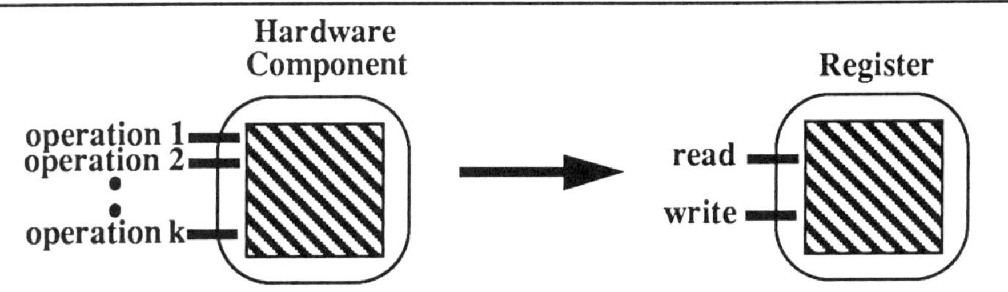

Figure 6.6 Using data abstraction for modeling hardware

As suggested above, either functional abstractions or data abstractions can be utilized to describe hardware or software. This idea is captured in the definition of a decomposition graph. The decomposition graph also incorporates the modeling concepts of abstraction level and interpretation level.

Definition 6.7 *Decomposition graph*

Referring to equations (6.1)-(6.4), a decomposition graph *DG* is a hierarchical, directed graph consisting of nodes *N* and edges *E*. The nodes can correspond to either functional abstractions or data abstractions. The edges represent "consists of" relationships between the nodes, reflecting the fact that a node can be expressed in terms of more primitive nodes. Thus, a decomposition graph employing functional abstractions depicts a functional decomposition, and a decomposition graph expressed in terms of data abstractions describes a data decomposition [250], a decomposition based on abstract data types. As can be observed, complexity management is an important consideration in this representation.

$$DG = (N,\ E) \tag{6.1}$$

$$N = \{n_1, n_2, \ldots, n_z\} \tag{6.2}$$

$$E \subseteq N \times N,\quad (n_i, n_j) \in E \leftrightarrow n_i \ \mathit{consists\ of}\ n_j \tag{6.3}$$

$$L = \{n_j |\ (n_i, n_j) \in E \ \mathit{and}\ (n_j, n_k) \notin E,\ \ k = 1 \ldots |N|\} \tag{6.4}$$

The leaf nodes *L* derived from the decomposition graph are used to define a virtual (abstract) machine with a corresponding virtual instruction set. In a functional decomposition, the leaf nodes comprise the virtual instruction set. In a data decomposition, the leaf nodes represent the most primitive abstract data types in the graph, and the operations to be performed on these data types constitute the virtual instruction set. Thus, a virtual instruction set can be derived through either functional decomposition or data decomposition.

To make these ideas more concrete, in the case of functional abstractions, the leaf nodes will take the form shown in equations (6.5)-(6.6). In the general case, a function f_j transforms a set of inputs X_j and state S_j to a set of outputs Y_j and a new state S_j'. In addition, a set of constraints C_j can be provided in terms of various metrics, such as performance, reliability, and cost. These constraints can be used to guide the derivation of an implementation for f_j. As indicated in equation (6.7), the leaf functions are used to derive a virtual instruction set V.

$$n_j = (f_j, X_j, Y_j, S_j, C_j) \tag{6.5}$$

$$f_j:(X_j,S_j) \rightarrow (Y_j,S_j') \tag{6.6}$$

$$V = \{v_1, v_2, \ldots, v_y\} = \bigcup_{j=1}^{|L|} f_j \tag{6.7}$$

With data abstractions, the leaf nodes will take the form shown in equation (6.8). Any node consists of a set of functions F_j (operations) defined on the abstract data type, with each function f_j expressed as in equation (6.6). As in the case of functional abstractions, a set of constraints may also be specified. Referring to equation (6.9), the set union of the operations taken over all $n_j \in L$ produces a virtual instruction set. As an aside, note that a specification of a functional abstraction is a degenerate case of a specification for a data abstraction.

$$n_j = (F_j, S_j, C_j) \tag{6.8}$$

$$V = \{v_1, v_2, \ldots, v_y\} = \bigcup_{j=1}^{|L|} F_j \tag{6.9}$$

The decomposition graph is a unified representation in several senses. First, the nodes can represent either functional abstractions or data abstractions. Second, the graph serves as a representation that can be used for both software and hardware. There is also a third possible element of unification, that of employing common modeling elements for hardware and software.

Note that the decomposition graph does not express all aspects of the decomposition. In particular, the algorithmic aspects of a decomposition are an implicit part of the representation. For example, in a functional decomposition, the algorithm that combines the lower level functions to produce the higher level function is not visible in the representation. Thus, the decomposition graph encapsulates the algorithmic aspects of a decomposition, regardless of whether a functional decomposition or a data decomposition is performed.

For the remainder of this chapter, decomposition graphs employing functional abstractions will be emphasized. Although not explicit in equations (6.1)-(6.4), the graph is a function of both level of abstraction and level of interpretation. The discussion below focuses on the different levels of interpretation that can be associated with the graph. In the notation, the special label *NULL* indicates that the corresponding parameter is undefined, that is, no semantics are associated with the parameter.

Given a decomposition graph based on functional abstractions, the virtual instructions can have different interpretation levels (see equations (6.10)-(6.11)). Assuming that the state has been defined, the extreme levels are *uninterpreted* and *interpreted.* Uninterpreted models ignore the identity of the functions as well as the corresponding inputs and outputs [112][90][91]. Interpreted models preserve the identity of these elements. Note that temporal information, in the form of delays, may be associated with each of these interpretation levels to allow for performance analysis.

Uninterpreted

$$v_j = (NULL, NULL, NULL, S_j, C_j) \tag{6.10}$$

Interpreted

$$v_j = (f_j, X_j, Y_j, S_j, C_j) \tag{6.11}$$

Interpreted models require some elaboration. There are two types of function interpretation: 1) specifying the function to be performed and 2) providing a functional description that produces an input-output transformation. The inclusion of condition 2) increases the detail within the model. However, the lack of condition 2) may result in undecidable conditions within the model and can lead to the test on value problem.

The test on value problem arises in the execution of a program on a processor model lacking functional transformations. Specifically, because the processor model lacks the ability to perform functional transformations, the internal state is left primarily undefined, aside from any initial state the processor may have. As a result, undecidable conditions may arise, such as testing the value of a register during a branch condition. Such a condition may be dealt with by employing, for example, probabilistic branching.

6.4 RELATED WORK

Compilers and synthesis capabilities employ several types of intermediate representations. Some common representations used in compilers include abstract syntax trees and tuples [187]. Hardware synthesis [12][152] and cosynthesis tools [199][216] utilize data/control flow representations and syntax trees. The ADAS environment [15] uses directed graph representations for hardware and software to support early performance evaluation.

The decomposition graph differs from these representations in several ways. The representations described above employ functional abstractions. The decomposition graph incorporates descriptions based on either functional abstractions or data abstractions. Compilers and synthesis capabilities utilize internal representations for software and hardware generation. The focus of the decomposition graph is to support the design of hardware and software elements. Finally, the decomposition graph brings together several common ideas associated with hardware and software, such as the modeling concepts of abstraction level and interpretation level.

6.5 SUMMARY

This chapter has presented several benefits of providing a unified representation for hardware and software. These benefits include the use of common design and analysis techniques with the possibility of applying techniques from one domain to the other (cross fertilization). Specifically, common analysis techniques can be utilized for the determination of performance, reliability, and formal verification of correctness. For example, techniques for analyzing bottlenecks due to critical computations can be applied in a uniform fashion. From a more qualitative standpoint, a unified representation provides a common modeling paradigm that can be understood by both hardware and software developers. All of these ideas serve to unify the hardware and software design processes.

A unified representation, based on decomposition graphs, was introduced. The decomposition graph is a unified representation in several senses. The nodes can represent either functional abstractions or data abstractions. Also, the graph serves as a representation that can be used for both software and hardware. The decomposition graph incorporates the modeling concepts of abstraction level and interpretation level, two aspects which are common to both hardware and software models.

Chapter 7
An Abstract Hardware/Software Model

This chapter presents an abstract hardware/software model [245] that utilizes a unified representation. The unified representation is based on functional abstractions and employs data/control flow concepts. The first section states the requirements and applications of the model. The next section introduces some existing models of hardware/software systems. Several ideas from these models have been incorporated into the abstract hardware/software model. The abstract hardware/software model is then described. This presentation is followed by a discussion of the model's implementation in ADEPT. An example is provided to illustrate various aspects of the model. The model's generality is discussed. The last section compares the model with other models that have been developed for hardware/software systems. The utility of the model is further demonstrated in Chapter 8.

7.1 REQUIREMENTS AND APPLICATIONS OF THE MODEL

The motivations for using an abstract hardware/software model, discussed in Chapter 1 and Chapter 5, impose certain requirements on the model. Specifically, the model should be executable, language-independent, and integrated. Language independence eliminates any biases introduced by languages used for hardware or software development. The term integrated implies the incorporation of the hardware and software descriptions of a system within a common simulation environment. A unified representation supports both of these ideas.

Referring to Figure 7.1, there are many applications of the abstract hardware/software model. The model can be utilized for general performance evaluation, identification of software bottlenecks, evaluation of hardware/software trade-offs, and evaluation of design alternatives. This chapter focuses on general performance evaluation. The remaining applications of the model are illustrated in the following chapter.

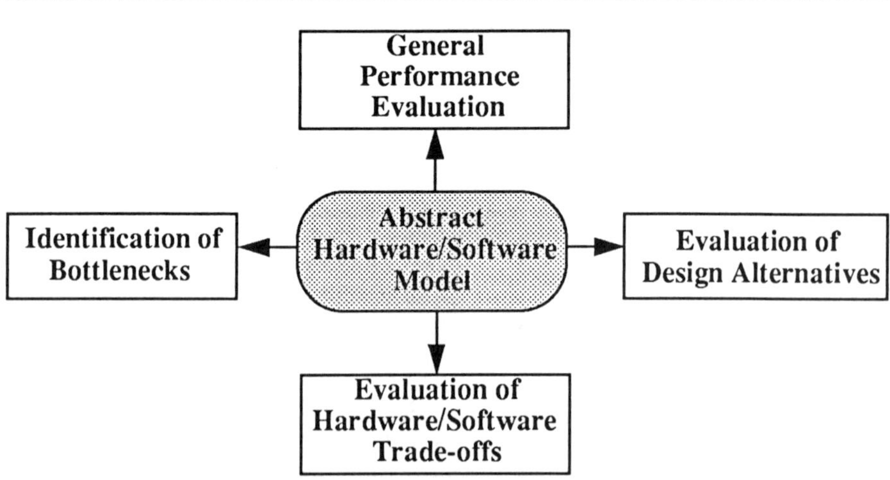

Figure 7.1 Applications of an abstract hardware/software model

7.2 MODELS OF HARDWARE/SOFTWARE SYSTEMS

In this section, two related models for hardware/software systems are presented. First, a model for interpretive systems (not to be confused with the modeling concept of interpretation presented earlier) is described. Next, a discussion of the request/resource model is provided.

7.2.1 Interpretive Systems

A *machine* [251][252] consists of three components: an *instruction set*, a *storage*, and an *interpretive mechanism*. The instruction set defines the set of primitive operations supported by the machine. A storage is composed of addressable locations which contain values. A *state* refers to a particular configuration of storage. The interpretive mechanism executes a program, consisting of *instructions* or *commands*, causing transformations in the machine's state.

The interpretive mechanism can also be a machine, referred to as the *host machine* (also called base machine), creating a hierarchy of machines. The higher level machine is called an *image machine* (also called target machine or virtual machine). The program that executes on the host machine is called an *interpreter program* or *emulator*. Such a machine hierarchy represents a one level interpreter. However, the image machine can also be a host machine for a higher level image machine, producing a two level interpreter. One can extend this idea to N levels. Therefore, hardware/software systems can be viewed in terms of one or more layers of interpreters. A fundamental aspect of an interpreter is its fetch/execute behavior, regardless of whether the interpreter is a hardware element, a microprogram, or a machine level program.

Hoffman [253] developed a classification of interpretive systems based on the number of interpreters (actually interpreter programs) present. In Hoffman's model, two types of interpreters are described. In a *command interpreter*, the commands, that is, the instructions to be interpreted, are not stored in a program memory. In a *program interpreter*, the instructions to be interpreted are part of a stored program. These two types of interpreters are illustrated in Figure 7.2.

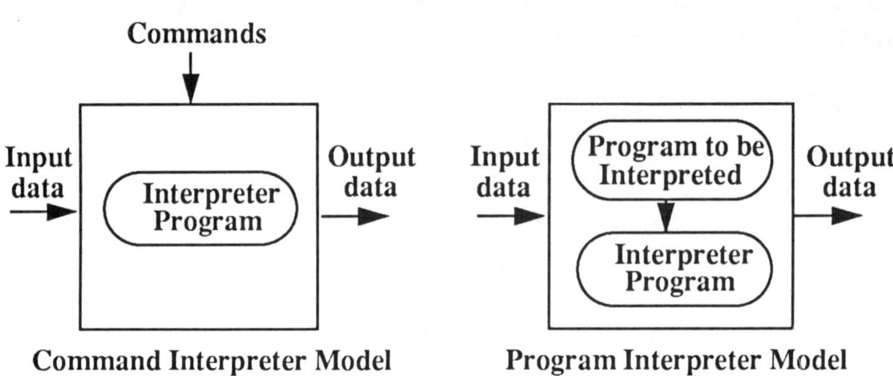

Figure 7.2 Command interpreter and program interpreter (From [253], © 1983. Reprinted by permission of Elsevier Science B. V.)

In the notation that follows, an arrow with a single line "$\rightarrow$" denotes "interpreted by", and an arrow with double lines "$\Rightarrow$" designates "executed by". The symbol "*P*" is a program to be interpreted. The symbol "*I*" corresponds to an interpreter. A software interpreter will be distinguished from a hardware interpreter by providing a superscript *s* or *h*, respectively. The symbol "*H*" refers to the physical devices within the system. Thus, a one level program interpreter can be represented as "$P \rightarrow I^s \Rightarrow H$" or as a triplet (P, I^s, H). Here, a program *P* is interpreted by executing a software interpreter I^s on hardware *H*. This representation corresponds to a machine level program being interpreted by a microcoded processor.

Beyond one level interpretive systems, several types of two level interpretive systems can be described using this model. Systems containing microprocessors that are implemented with microcode and nanocode, such as the MC68020, are one example. Other examples include systems which interpret intermediate representations, for example, P-code [254] and U-code [255]. P-code is an intermediate code for Pascal programs, and U-code is an extension of P-code which consists of instructions for an abstract stack machine. A more detailed discussion on software interpreters for intermediate representations can be found in [256].

Using the notation described earlier and assuming that a microcoded processor is utilized, interpretive systems for intermediate representations can be described as a four-tuple (P, I_1^s, I_2^s, H). In this description, P is the intermediate representation being interpreted. The two software interpreters, I_1^s and I_2^s, correspond to a machine level language interpreter for P and a microded interpreter for the machine language, respectively.

An assumption of this model is that the levels strictly build on each other. In some circumstances, a given level can be interpreted by several lower levels. Models that attempt to accommodate for these systems have also been developed [253][257].

Given this model of interpretive systems, a number of design problems can be investigated based on what portions of the interpretive system are fixed or variable. For example, if only the hardware resources and the microinstructions are fixed within a processor, several possible microprograms can be written to provide various types of functionality, such as different instruction sets. If all levels below the microprogram are fixed, the functionality of a processor can only be modified by writing different machine level programs. Thus, the portions of the interpretive system that are fixed influence the allowable hardware/software trade-offs that can be performed.

7.2.2 The Request/Resource Model

The request/resource model [258] has also been utilized in descriptions of hardware/software systems. This model has been used as a basis for descriptions of computer organizations. In this model, a system consists of a *requestor* and a *server*. The requestor is a program, and the server is a set of resources. The distinction between a program and a resource is somewhat fuzzy. In some circumstances, a resource can be another program and vice versa.

Referring to Figure 7.3, a program *P* is defined in terms of a sequence (stream) of requests (tasks) for resources, whose purpose is to process the requests. Since a request can also be considered a program, each request can be expressed in terms of subrequests, producing a request hierarchy. The service for a request is provided by a subtree structure consisting of a hierarchy of subrequests that terminate in the physical resources of the system. Time is associated with these leaf nodes, corresponding to a physical activity, such as adding two registers and writing the result to a third register. The satisfaction of a request is subject to the resource's availability.

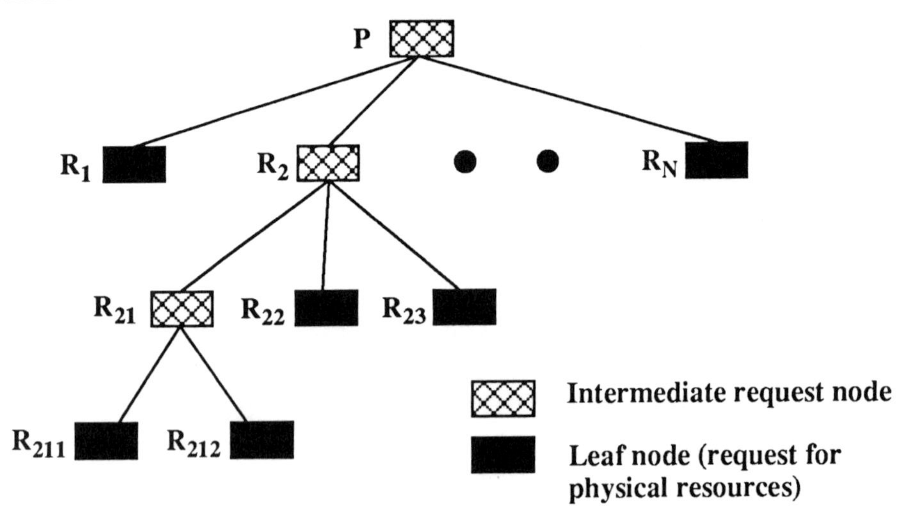

Figure 7.3 Service hierarchy (From [258], © 1972 IEEE)

An abstract request/resource model has been used by several system level modeling capabilities to support hardware/software analysis early in the design process. For example, in SES/Workbench [76] and ADAS [259], software operations request resources and may "block" if a hardware resource is not available. Models based on this paradigm are useful for analyzing the effects of concurrency and resource contention on performance.

7.3 AN ABSTRACT HARDWARE/SOFTWARE MODEL

This section presents the abstract hardware/software model. An informal view of the model is first provided. A model employing a unified representation based on functional abstractions is then formally described.

7.3.1 An Informal View of the Model

Instruction set level representations (see Figure 7.4 (a)) are the most familiar to those who work with hardware/software systems. In this representation, the software consists of a sequence of instructions which are fetched and executed by a processor. Of course, other hardware devices may be present, such as ASICs, along with other circuitry, for example, miscellaneous glue logic. Regardless of the type of hardware present, the models of these components are complex descriptions with detailed functionality and temporal information.

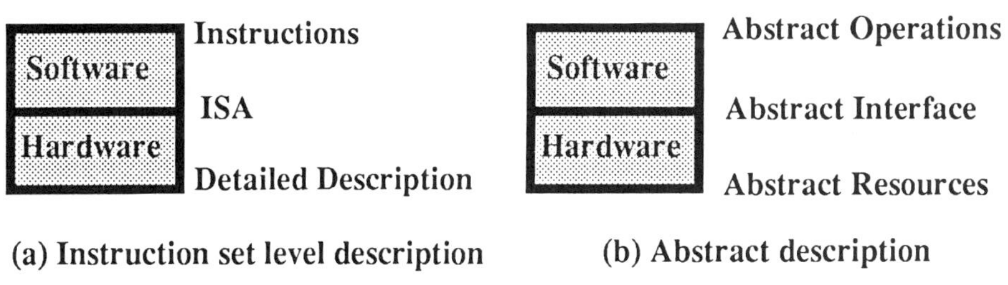

Figure 7.4 Instruction set level and associated abstract description

The relationship between software and hardware can be expressed as one in which operations in the software utilize resources within the hardware and manipulate its state. For example, the execution of an instruction *ADD R1, R2* within a load-store machine utilizes, among other resources, the register file and the arithmetic logic unit. In addition, a portion of the processor state, which includes the condition codes and the register file contents, is updated.

Because of the detail present in instruction set level descriptions, it is desirable to investigate abstractions of the form shown in Figure 7.4 (b). The purpose of this naive depiction is to convey the "flavor" of an abstract hardware/software model. Unified representations based on data/control flow concepts are used to describe the software and hardware. The abstract interface depends on the primitive operations provided in the software and those supported by the underlying hardware. In general, these operations may or may not correspond directly to those found in an ISA level model. The resources can be uninterpreted, lacking the ability to perform functional transformations.

An important modeling consideration is the amount of detail required within a hardware/software model in terms of both level of abstraction and level of interpretation. In developing a model, the detail present depends on what attributes are of interest and how the model is to be used. If an absolute comparison of execution time between two algorithms executing on different processors was required, very detailed descriptions of both hardware and software would be necessary. The software would have to be described at the instruction set level. The hardware models would consist of functional descriptions and temporal information that take into consideration caching, pipelining, number of functional units, along with any other inherent concurrency. However, in some circumstances, particularly during the early stages of the design process, a gross performance evaluation may be sufficient, allowing some details to be ignored.

7.3.2 A Model Based on Functional Abstractions

Definition 7.1 *Hardware/software model*

As indicated in equation (7.1), an abstract hardware/software model *HSM* consists of a software model *SM* and a hardware model *HM,* each of which can be described using the unified representation developed in Chapter 6. Specifically, each model can be expressed as a decomposition graph employing functional abstractions. This *HSM* description represents a single software program executing on a single processor. However, the definition can be extended to accommodate more complex systems.

$$HSM(l^s, l^h) = (SM(l^s),\ HM(l^h)) \qquad (7.1)$$

In the notation above, the symbols "l^s" and "l^h" encapsulate both level of abstraction and level of interpretation, (l^a, l^i), for software and hardware, respectively. In order to execute a *HSM*, the abstraction levels of the *SM* and *HM* must be *consistent.* By consistent, it is meant that the *SM* has a set of primitive operations that are "understood" by the *HM*.

A software model and a hardware model can have different levels of interpretation. Given consistent software and hardware models with different interpretation levels, it is possible to determine whether the combined hardware/software model can be used for performance analysis, functional verification, or both. Of course, other types of analyses can be performed. However, performance analysis and functional verification will be focused on.

The table in Figure 7.5 illustrates the different types of analyses that can be performed with a hardware/software model. In this example, it is assumed that delays, if present, are contained only in the *HM*. The table entries are labelled as "*P*", "*F*", "*B*", or "*N*" depending on whether the *HSM* can be used for performance analysis, functional verification, both, or neither, respectively. For example, the execution of an interpreted *SM* on an interpreted *HM* with no delay does not allow performance analysis but does allow functional verification. Performance analysis is not possible since the model contains no temporal information. Functional verification of the hardware/software model is supported because the state of the model is maintained. As a result, no undecidable conditions exist in the model.

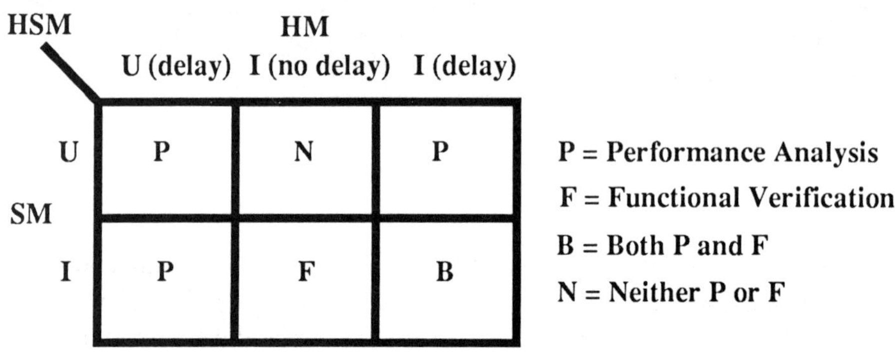

Figure 7.5 Some possible analyses with a hardware/software model

Some models are more meaningful than others. Of course, deciding whether a model is meaningful or not is somewhat subjective and is influenced by the user's objectives. As an example, an interpreted software model can be executed on an uninterpreted hardware model. However, the results derived from the execution of such a hardware/software model will probably not accurately reflect those obtained from the final hardware/software implementation. To more accurately reflect the system being developed, a hardware/software model with more detail would be required. Ultimately, the only way to provide a hardware/software model for both functional verification and performance analysis purposes is to utilize an interpreted software model and an interpreted hardware model with temporal information.

7.4 MODEL IMPLEMENTATION IN ADEPT

There are several possible implementations of the abstract hardware/software model. In the ADEPT environment, the structure in Figure 7.6 was used to develop abstract models based on structured programming concepts [112][119]. The structure consists of a software model (**SM**) and a hardware model (**HM**), each of which can be described using data and/or control flow graphs. Thus, each model can be a data flow graph, a control flow graph, or a combination of data and control flow. Embedded in this hardware/software model are the ideas of interpretive systems and the abstract request/resource paradigm of software functions requesting and then utilizing hardware resources.

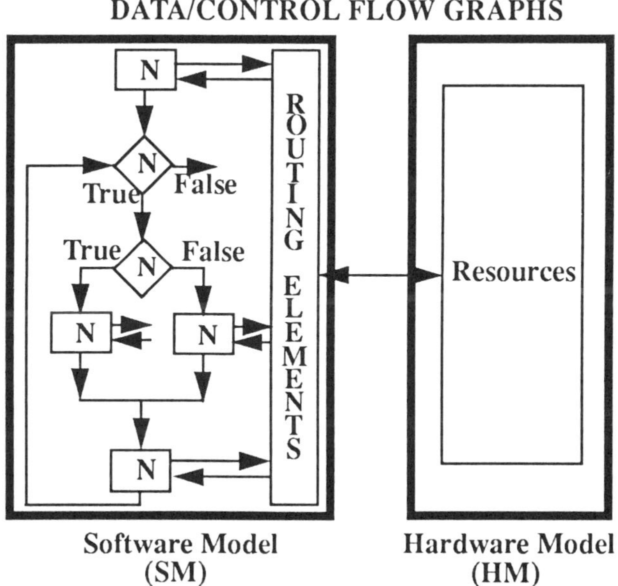

Figure 7.6 Structure of an abstract hardware/software model

The software model is a decomposition graph in which the individual operations (or nodes, **N**) are expressed in terms of their resource requirements. This description can be provided in several forms, such as three-address code representations used in compilers [260]. The hardware model is a decomposition graph that consists of the abstract resources required by the software operations, such as functional units, registers, or local memory. Software execution involves the request and subsequent utilization of the hardware resources over time.

7.4.1 The Software Model

A software model can contain process nodes (function nodes), predicate nodes (decision or branch nodes), collector nodes, and delay nodes. The first three nodes can be used to describe a variety of control structures in structured programming approaches [119], such as if-then-else, while-do, and case. Process nodes represent functional transformations, such as store or multiply. In general, the granularity of the node can be arbitrary. These nodes have a single input line and a single output line. Each line corresponds to control flow. Predicate nodes direct execution control based on whether an expression evaluates to true or false. Therefore, predicate nodes have a single input line and two output lines. Examples of these nodes include “less than”, “greater than”, and “equal to”. Collector nodes combine two or more input lines into a single output line. The delay node can be considered an uninterpreted process node.

Figure 7.7 illustrates some ADEPT software modules which have been developed to implement the nodes described above. The names of the modules appear in parentheses. Tokens flow along the input and output lines (ports). It should be noted that in some cases an “extra” output port is present in the software module. This extra output is necessary to permit execution of the node by the hardware model. For example, in the floating point multiply (*FMPY*) software node shown in Figure 7.7 (a), one input *in_1* and two outputs, *out_1* and *out_2*, are present. The ports *in_1* and *out_2* are used for control flow as in a typical process node. The output *out_1* is employed to send a request for node execution (in the form of a token) to the hardware model.

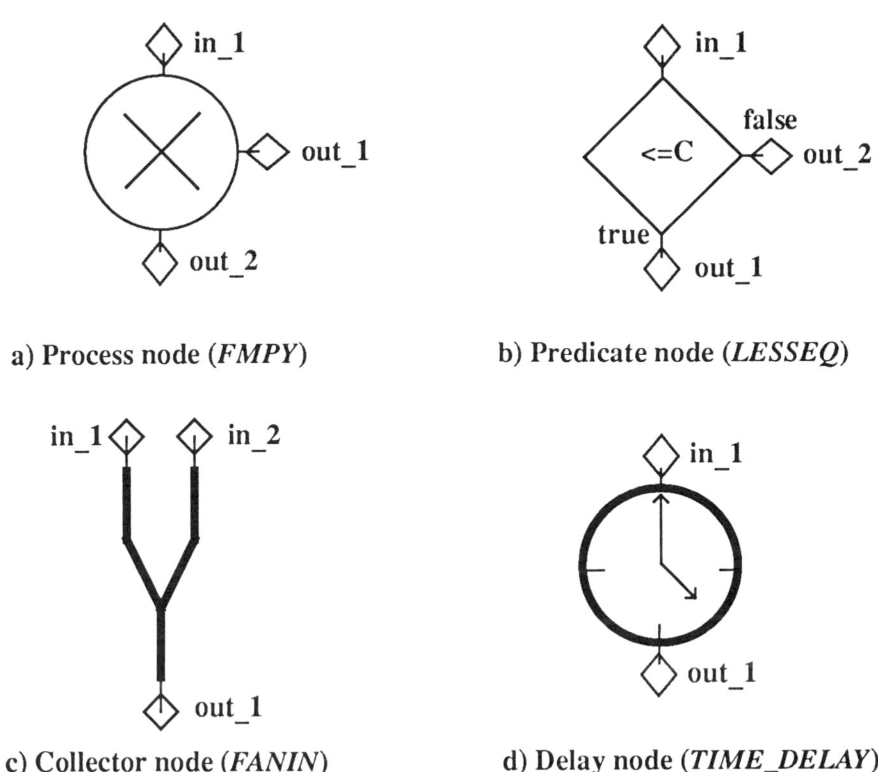

a) Process node (*FMPY*)

b) Predicate node (*LESSEQ*)

c) Collector node (*FANIN*)

d) Delay node (*TIME_DELAY*)

Figure 7.7 Some example modules used in ADEPT software model

The remaining modules in Figure 7.7 have the same number of inputs and outputs as the nodes described above. The *LESSEQ* module routes an incoming token to either the true or the false output based on the result of the comparison. Although not shown, a predicate node which allows probabilistic branching has also been developed. The *FANIN* module is used to merge token flows from two different sources, such as the true and false paths of a predicate node. This module is utilized in loops as well. Note that the *TIME_DELAY* module, which is an example of a delay node, can be viewed as a process node in which a delay has been associated with the computation, but no function has been specified.

Because process and predicate nodes represent computations, these nodes are the only ones that can be executed by a hardware model. Therefore, only these nodes may contain extra output ports. However, in some circumstances, process and predicate nodes are used solely for controlling the software model's execution. As a result, the nodes are not executed by the hardware model.

Delay nodes are used for abstracting a computation as a single, aggregate time delay. These nodes manage complexity and allow a model to be described at different levels of detail, permitting one to focus on those aspects of interest. In addition, delay nodes can be utilized to establish timing constraints. The node can also be refined after specifying its function.

The modules used to implement process and predicate nodes are described as VHDL processes. Generics are associated with these elements which allow parameters to be passed down into the VHDL code. For those software nodes which are executed by a hardware model, these parameters include resource information, for example, where source operands are to obtained from and where results are to be written (register, memory, and so on).

Although these ADEPT software modules are described in terms of VHDL, their abstract behavior is equivalent to an interconnection of existing ADEPT modules. This equivalence to existing modules allows the corresponding Petri net descriptions to be extracted and utilized if so desired, although a simpler Petri net representation is possible by converting the control flow graph to a Petri net directly.

As an example (see Figure 7.8), an executable process node is equivalent to a collection of *SC_D/CONSTANT* pairs connected to a *SEQUENCE* module. Each *SC_D/CONSTANT* pair colors a specific tag field of the incoming token with an integer value which designates resource information. In a three address code representation, four of these pairs would be required. One pair would specify the operation to be performed (*func*), while three pairs would be utilized to specify the destination (*dest*), source of the first operand (*src1*), and source of the second operand (*src2*). Thus, tag fields 1 through 4 of the incoming token (arriving at *in_1)* would be colored (populated) as shown in equation (7.2). Upon arriving at the input of the *SEQUENCE* module, this *request* token would be sent to the hardware model through *out_1*. Once accepted by the hardware model, a token would be placed on *out_2*, enabling the next node for execution. A more detailed discussion of token coloring and node execution is provided later in the chapter using an example.

$$token = (func, dest, src1, src2) \tag{7.2}$$

Process nodes, predicate nodes, and collector nodes can be used to construct several basic control structures [112][119], such as if-then-else and while-do. A subset of these control structures can serve as a basis for structured programs [261]. Before discussing how the ADEPT modules can be utilized to implement these constructs, it is worthwhile to elaborate on the concept of a structured program. The discussion here is found in [112].

A *proper program* is one which satisfies the following two conditions. The program's control structure has a single entry line and a single exit line (representing control flow). In addition, for every node within the program, a path through that node exists from the entry line to the exit line. The implication of the latter condition is that there are no infinite loops and no unreachable nodes. A proper program can have parts that are also proper. These parts are called proper subprograms.

A *prime program* is a proper program in which every proper

subprogram has at most one node. Several control structures can be viewed as prime programs. As an example, the sequence (concatenated process nodes), if-then-else, and while-do structures are all prime programs. Other control structures that are also considered prime programs include if-then and do-until.

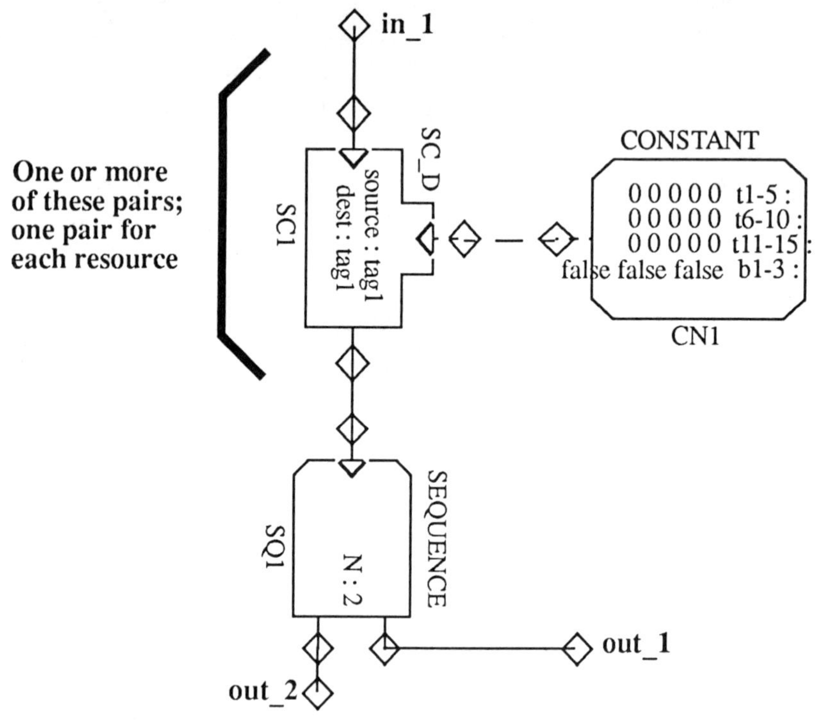

Figure 7.8 Equivalent ADEPT module description of a process node

A *compound program* is defined as a program obtained by replacing process nodes within a prime program by another prime program. As a special case, prime programs are considered compound programs. The prime programs employed can be restricted to a basis set.

For example, one basis set is {sequence, if-then-else}. Another basis set is {sequence, if-then-else, while-do}. A *structured program* can then be defined as a compound program derived from a fixed, basis set of prime programs.

Referring to Figure 7.9, the ADEPT software modules can be utilized to describe the sequence, if-then-else, and while-do control structures (as well as others). For simplicity, it has been assumed below that only the process nodes are executed by the hardware model. The arrows correspond to entry and exit lines. The rectangle in the if-then-else construct is another type of collector node (*UNION* module), and the rectangle in the while-do construct is an ADEPT "double buffer" module, which ensures proper token flow.

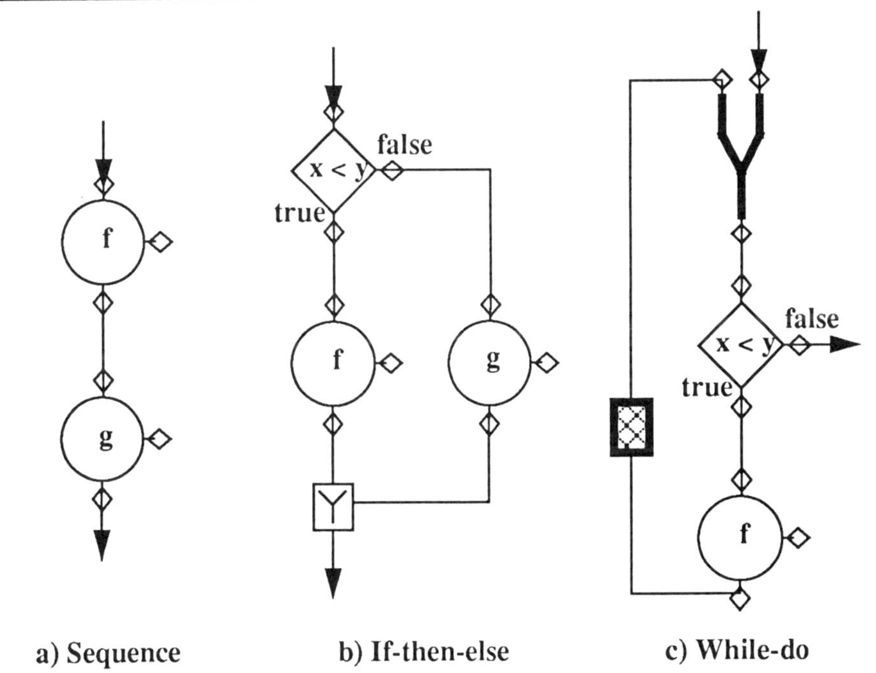

Figure 7.9 ADEPT versions of sequence, if-then-else, and while-do

7.4.2 The Hardware Model

In order to execute the operations in the software model, a processor model is required. The hardware model can be described in any number of ways. However, because of the interpretive nature of processors, it is natural to model the hardware in terms of interconnected fetch and execute units. In some circumstances, multiple execute units may be employed, allowing more than one operation to be executed concurrently. The advantage of using a fetch/execute structure will become apparent shortly.

A hardware model can contain resource nodes, delay nodes, predicate nodes, and collector nodes. The resource nodes correspond to functional units, such as an ALU, processors, or memory elements. Some examples of resource nodes include servers with and without queues. Delay nodes can also be utilized to represent abstract resources. For example, a delay node may be used to model a fetch process, indicating that the memory is busy for a period of time specified by the delay. Resource delays are typically provided in terms of clock cycles. Through generics, the user can indicate the clock cycle time and the number of cycles required to perform a particular operation. A resource with a queue (*RESOURCE*) and an abstract functional unit (*ARITHMETIC_UNIT*) are shown in Figure 7.10. One application of predicate nodes is the decoding of an operation to be executed by one of several units. Collector nodes are used to merge token flows.

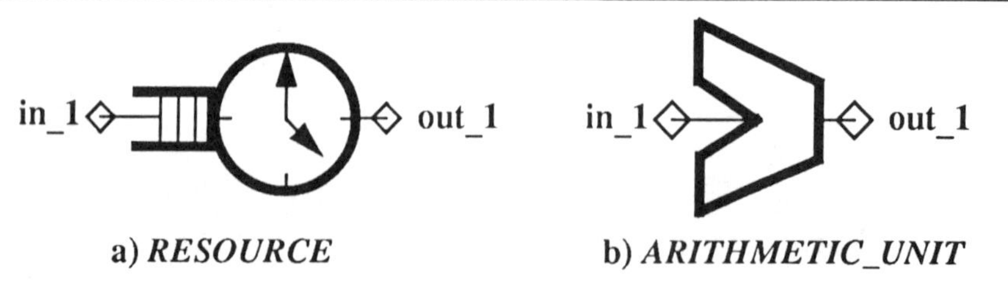

a) ***RESOURCE*** b) ***ARITHMETIC_UNIT***

Figure 7.10 Some resource nodes used in ADEPT hardware model

7.4.3 The Effect of State on Model Execution

An issue that has been ignored up until now is the effect of state on the execution of hardware/software models. In general, the deterministic execution of a software model containing loops and branches requires that state be examined. If no state is available, probabilistic branching is necessary to address the test on value problem. In the ADEPT environment, state can be captured in primarily two ways. The first is by storing state in the circulating tokens as the model executes. The second technique is to use an explicit memory in the hardware model. Although queues are available in the ADEPT environment to store state, these modules are first-in-first-out (FIFO) elements, and no convenient indexing mechanism exists for reading and writing arbitrary locations.

The technique of storing state within a circulating token is useful if the number of "variables" is small. For example, this technique is practical if a collection of operations within a loop is to be executed a fixed number of times. In other words, the number of loop iterations is known statically (at compile-time). In this situation, the iteration number can be kept in a specific tag field of the token, which can be manipulated by increment operations and examined by any branch operations.

A more general solution is to use a memory element within the hardware model which stores the state of the software computation as it proceeds. A drawback of this approach is that predicate nodes must obtain information regarding the result of the operation, that is, whether the condition was true or false, from the hardware model. Only upon receiving this result can the node route a token along either of these paths. The transmission of this information necessitates the explicit inclusion of interconnections from the hardware model to the software model. These interconnections can complicate the model. Because of this reason, a bidirectional arrow is provided between the software model and the hardware model (see Figure 7.6). Information flow from the software model to the hardware model is in the form of a request for computation, and information flow in the opposite direction indicates the return of information to a software node. In some cases, a completion signal may be returned to the software model.

Assuming a graph-based representation for the software model, Figure 7.11 illustrates a modeling structure which utilizes a memory element to maintain state. The interconnection problem can be eliminated by not using a graph-based software representation. However, other types of analyses, such as bottleneck analysis, and the ability to perform trade-offs become more difficult.

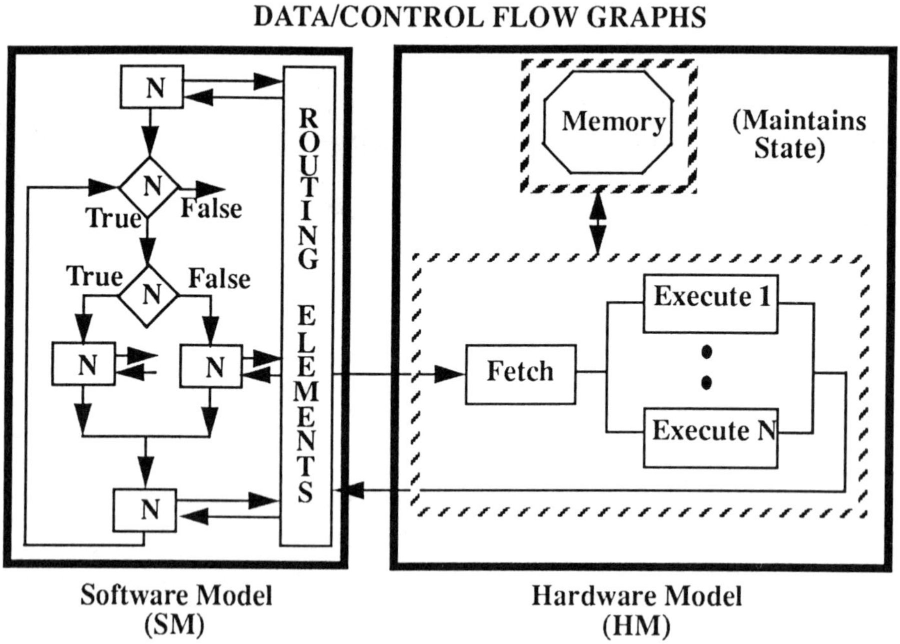

Figure 7.11 A modeling approach employing a memory element

The modeling structure in Figure 7.6 is quite general. The structure can be utilized to represent software execution on a processor, or a special purpose hardware element consisting of microcode activating a set of hardware resources. Also, because the fetch/execute behavior is common to both hardware and software interpreters, this modeling structure is useful for representing and analyzing several types of interpretive systems. This structure also allows the hardware model to be refined by incorporating lower level implementations. Therefore, the model supports hybrid modeling [21].

7.5 AN EXAMPLE

A *HSM* consisting of a finite impulse response (FIR) algorithm executing on an abstract digital signal processor (DSP) is presented in this section. A control flow graph representation of the FIR algorithm in ADEPT, containing parallel multiply-add (x || +) and floating point add operations (+), is shown in Figure 7.12. The circular symbols correspond to process nodes, and the diamond-shaped symbol represents a predicate node (<= c). The other nodes are used to control token flow.

Execution of the *HSM* is initiated by the arrival of a token at the topmost process node. This node sets a field within the token to an integer, corresponding to an initial value for the number of loop iterations. The Y-shaped icon below the process node transfers either the initialized token or a token with a newly incremented value to the predicate node. The predicate node routes the token along either the true path (bottom arc) or the false path (right arc). The predicate node, multiply-add node, and the increment node (+1) are part of a loop that is executed *N* times, where *N* is a user-specified parameter corresponding to the length of the filter. The rectangular icon in the feedback path of the graph ensures that the tokens flow properly as the model executes.

In this example, only the multiply-add and floating point add operations are "executed" by the DSP. The decision node and the increment node are being used solely to model the process of iteration and thus, ensure that the multiply-add operation is executed N times. Therefore, the cost associated with loop overhead is not being taken into account.

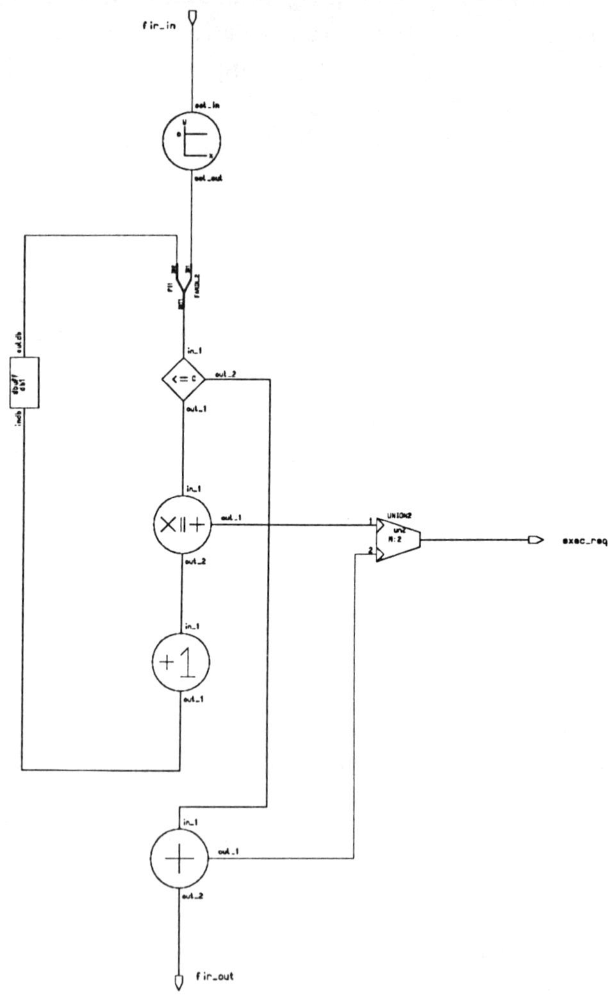

Figure 7.12 FIR software model

Execution of a software node occurs in three steps. In the first step, the arrival of a token at the top arc of a node enables the node for execution. Each software node to be executed "colors" (populates) the token with the node's resource requirements as shown below. In this example, a floating point add operation (*FLTADD*) is to be performed on two register operands (*REG*), with the result to be written to a memory location (*MEM*).

$$token = (func, dest, src1, src2) \tag{7.3}$$

$$token = (FLTADD, MEM, REG, REG) \tag{7.4}$$

This request token is then sent to the DSP model via the node's right arc through a *UNION* module (far right), the sole routing element in this software model (see Figure 7.6). The *UNION* module places a token on its output (*exec_req* port) when a token appears on any one of the two inputs. Once this output token has been fetched by the processor, a token is sent to the node's bottom arc, enabling the next node.

The two DSPs employed in this example are the TMS320C30 and the TMS320C30-40 which have single-cycle execution times of 60 ns and 50 ns, respectively. There are several features of these processors which make them particularly suited for executing filter algorithms, such as circular addressing and parallel multiply-add operations. In addition, pipelined operation provides high throughput.

The DSP model consists of three portions: a fetch stage, a decode stage, and an execute stage. The operand fetch stage has been neglected for simplicity but can also be included to more accurately characterize the software performance. Note that the processor model acts as an interpreter, fetching and executing operations. Thus, this modeling structure can also be used to represent and analyze software interpreters as well.

An abstract model of the fetch stage is shown in Figure 7.13. The model accepts a software operation (request token) from the left of the figure via the *fetch_in* port, waits for a user-specified fetch delay, and passes the operation to the decode stage (not shown) via the *fetch_out* port. The *BUFFER* module (far right) sends the software operation to the decode stage and also allows a new incoming token to be received by the fetch stage, allowing pipelined behavior. Upon arriving at the execute stage, the request is either granted or blocked until a resource is available.

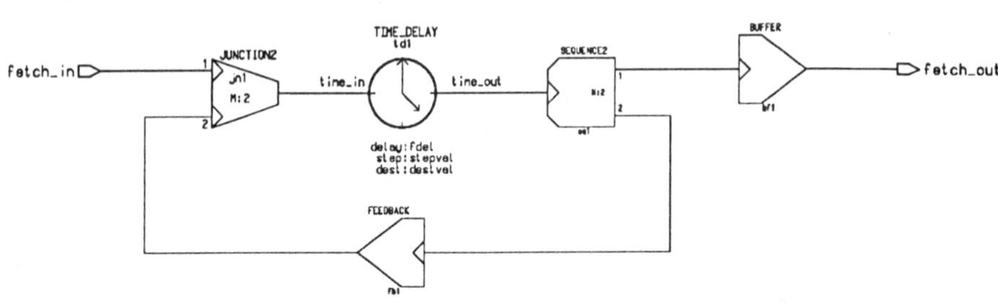

Figure 7.13 Fetch stage of digital signal processor

The execute stages of both processor models contain abstract resources, modeled as uninterpreted elements lacking function. Referring to Figure 7.14, the digital signal processor includes two abstract resources: a multiplier (center top) and an ALU (center bottom). Both of these resources are modeled as delay elements which do not perform any functional transformations on the input operands. In this model, a request token arrives at left from the decode stage (not shown). Multiply operations are routed to the top resource, and ALU operations are routed to the bottom resource. Parallel multiply-add operations utilize both resources. Once the operations have been "executed", the request tokens are consumed by the *SINK* module (far right) at which time the resources are freed. The remaining ADEPT modules are used for controlling token flow.

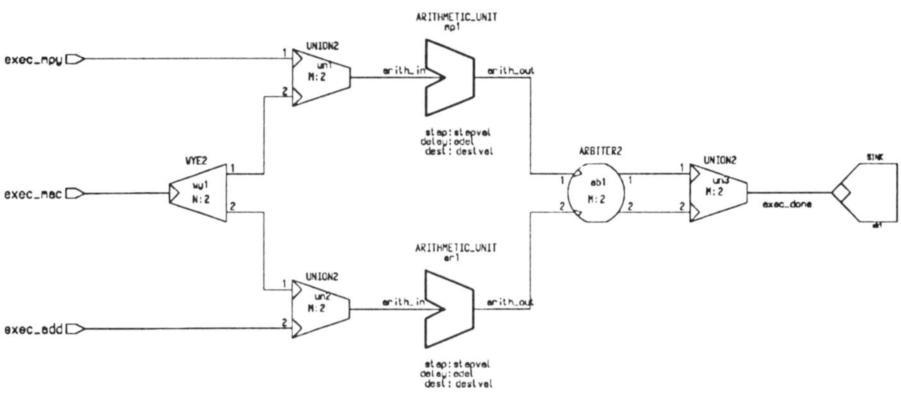

Figure 7.14 Execute stage of digital signal processor

The software model of the FIR can be executed on the DSP model to evaluate gross performance. Figure 7.15 contains the simulation results for the execution of several FIR filters on the TMS320C30 and TMS320C30-40 processor models. The execution times in microseconds are displayed versus the filter length. In the simulation, it was assumed that the fetch and execution of the operations each require a single clock cycle. It was also assumed that the fetch and execute stages operate in pipelined fashion. Because the operand fetch stage has been neglected for simplicity, the execution times do not reflect any latencies due to memory reads or writes for operands and results, respectively. However, this detail can be added if so desired.

In a similar fashion, Figure 7.15 displays the simulation results for the execution of various FIR filters on the MC68020/68881 processor models with different clock rates. The time to fetch operations was assumed to be three clock cycles, indicating that all memory reads were being performed from an external memory. The execution times for all operations were extracted from the MC68020/68881 data books. The software model contained only the floating point operations associated with the FIR since floating point performance was the aspect of interest. As a result, no concurrent execution was exploited.

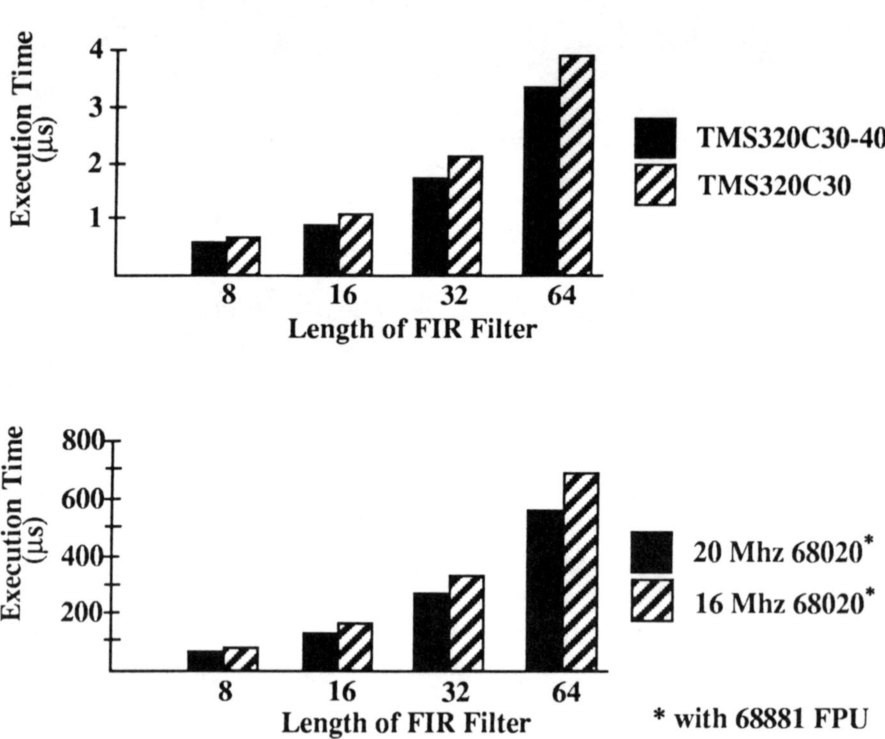

Figure 7.15 Simulation results for FIR

7.6 GENERALITY OF THE MODEL

In addition to the unified representation, there is another unifying concept hidden within the abstract hardware/software model. Specifically, the notion of an interpreter is common to both hardware and software. This concept is reflected in the fetch/execute structure of the hardware model. More generally, the hardware model can be considered a virtual machine abstraction *VM*, which may correspond to a processor or possibly a software program (see equation (7.5)).

$$M(I^S, I^V) = (SM(I^S),\ VM(I^V)) \qquad (7.5)$$

There are several observations that can be made regarding the virtual machine abstraction. The virtual machine provides a form of information hiding. The implementation may be a processor with a microcoded control unit or a hardwired control unit. However, this implementation information is hidden from the user of the virtual machine. As a result, complexity management is promoted.

Referring to Figure 7.16, based on the model of interpretive systems, a uniform view of hardware/software systems can be constructed, one that expresses interactions between software developers or between software and hardware developers. This generalized model consists of three fundamental elements: a *program to be interpreted*, an *interpreter*, and an *interface* between the program to be interpreted and the interpreter. The interface serves as a "contract" between one who develops the interpretable description and one who develops the interpreter (virtual machine).

As an example, the interface can be the ISA for a particular machine. This ISA serves as a specification that allows programmers developing machine language programs and compiler developers to write their respective programs with confidence. The ISA also guides the hardware designers who develop the machine. The machine can be constructed with a microcoded or a hardwired control unit as long as the requirements specified by the ISA are met.

In addition to serving as an interface between software developers and hardware developers, the same type of interface can be utilized between software developers. For example, the interpretable program may be an intermediate program to be interpreted. In this scenario, the interface would consist of the intermediate instruction set. In a manner similar to the ISA, this intermediate instruction set serves as a specification for the developer of a software interpreter.

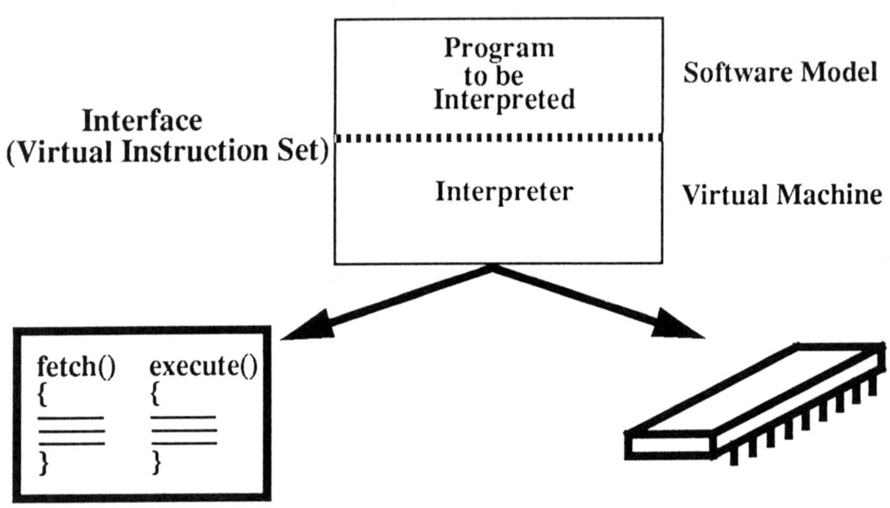

Figure 7.16 Software/software and software/hardware interactions

7.7 Related Work

There are several works which are related to the abstract hardware/ software model. The software model is influenced by ideas from several places. The computation structure [77][83] employs a data flow graph and a control flow graph to represent a computation. The control flow graph is utilized to analyze the performance properties of programs. Chu [93] uses control flow graphs with probabilistic branches for performance analysis. The concepts of process nodes, predicate nodes, and collector nodes are derived from structured programming techniques [112][119]. The notion of interpretation level within programs is present in works by Linger et al. [112], Peterson [90], and Kavi et al. [91]. Three-address code representations are found in discussions on compilers [187][260].

The hardware model is also influenced by several efforts. The concept of interpretation level, and particularly the combining of uninterpreted and interpreted models, is derived from Aylor et al. [21] and Auletta [92]. Hoffman's model of interpretive systems [253] is the primary idea behind the fetch/execute structure.

By incorporating hardware and software descriptions within a common simulation environment, the abstract hardware/software model captures the cooperative nature of hardware/software design. This distinguishing feature of the model sets it apart from many of the software analysis techniques described above and existing software estimation techniques [207]. However, it should be noted that the model can also be used for only software performance analysis as well as hardware performance analysis.

The abstract hardware/software model shares much in common with the model used by ADAS [15]. Specifically, both models attempt to address early hardware/software evaluation using abstract models with unified representations. Smith [63] discusses techniques for converting timed computation graphs, which are extensions of basic computation graphs [262], to augmented timed Petri nets for performance analysis. In a similar manner, the abstract hardware/software model can be converted into a Petri net for analysis purposes, particularly performance evaluation. Also, both models utilize the idea of software operations requesting hardware resources in some form. The concept of interpretation level is present as well.

However, there are some distinguishing characteristics. The abstract hardware/software model presented in this chapter incorporates three fundamental ideas: virtual machines, interpretive systems, and the request/resource model. Thus, the abstract hardware/software model can potentially be used to represent multilevel interpretive systems. The first two ideas are not inherent to the ADAS model. The abstract hardware/software model can also potentially be used as an integrated substrate during hardware/software development.

7.8 SUMMARY

This chapter has presented an abstract hardware/software model employing a unified representation. The unified representation is based on functional abstractions and utilizes data/control flow concepts. Specifically, a common modeling structure, that of hierarchical, directed graphs, is used to describe the software and hardware models. Such a language-independent representation provides a common modeling paradigm for hardware and software engineers. This approach permits hardware and software to be developed cooperatively in a common simulation environment.

An abstract hardware/software model has several applications. The model supports early evaluation, allowing the consequences of hardware/software decisions to be assessed before committing to a particular implementation. In addition to general performance evaluation, the model can be used to identify software bottlenecks, evaluate hardware/software trade-offs, and evaluate design alternatives. Designers can experiment with different software algorithms or vary the number and speed of hardware resources. The model supports evaluation at different levels of detail. This capability provides a designer the flexibility of focusing on those aspects of interest. The integration of both hardware and software descriptions within a common environment supports combined performance and reliability estimation at (possibly) several stages of the design process.

Hardware/software evaluations may be used to derive specifications for later design steps. For example, performance specifications can be utilized to guide the design of software and hardware components. Lower level implementations can also be incorporated into the model, supporting model continuity.

By viewing the hardware model as a virtual machine abstraction, the abstract hardware/software model can be generalized to represent interactions between software developers or between software and hardware developers. The interface between these development groups constitutes a contract between the users of the virtual machine and those who implement the machine. In both cases, the virtual machine hides design decisions. Also, the virtual machine provides a "substrate" which can be programmed.

Chapter 8

Performance Evaluation

This chapter can be considered an extension of Chapter 7. The application of the abstract hardware/software model is further illustrated using ADEPT. The first section elaborates on the applications of the model, which include general performance evaluation, the identification of software bottlenecks, the evaluation of hardware/software trade-offs, and the evaluation of design alternatives. Another potential use of the model, that of supporting the idea of an integrated substrate, is also discussed. Chapter 7 showed how the model could be utilized for general performance evaluation. In the second section of this chapter, the other applications of the model are demonstrated via several examples.

8.1 Applications of the Abstract Hw/Sw Model

This section expands on discussions in the previous chapter regarding the application of the abstract hardware/software model. In addition, the use of the model as a means of supporting the notion of an integrated substrate is discussed.

8.1.1 General Performance Evaluation

The abstract hardware/software model can be used as a representation for a software unit. As demonstrated in the previous chapter, the model may be used to assess the performance of several types of software units. Different implementations of a software function can be evaluated on the same hardware model. The effects of concurrency and speed of resources within a hardware model can be observed on the overall performance of the hardware/software system under consideration. The virtual machine approach allows a range of design alternatives to be evaluated, from general purpose to application specific.

8.1.2 Identifying Software Bottlenecks

Improving the execution time of critical computations can dramatically increase overall performance. The identification of software bottlenecks can reveal functions which require enhancement, either in the form of more efficient algorithms or hardware support. These techniques are applicable to both hardware and software elements. As shown in equation (8.1), the operator sensitivity metric S_i [77] can be used to quantify the portion of the overall execution time T taken by a software node i, where a node can be a single operation or an entire module. Software nodes with high sensitivity values are good candidates for improvement.

$$S_i = \frac{T_i}{T} \tag{8.1}$$

In the ADEPT environment, the calculation of an operator sensitivity is performed with the aid of the *MONITOR* module. As indicated in Figure 8.1, this ADEPT module is similar to a “voltmeter”, allowing the designer to probe the inputs and outputs of a node. A node may consist of one or more ADEPT modules. As the simulation proceeds, the *MONITOR* generates a file which contains the input-output latencies of the node being probed. The *FILE_WRITE* modules, shown at left and right, write the start and end times of the computation, respectively, into another file. Using the information contained in these two files, a post-processing program called *opsens* is then used to determine the operator sensitivities for the nodes of interest.

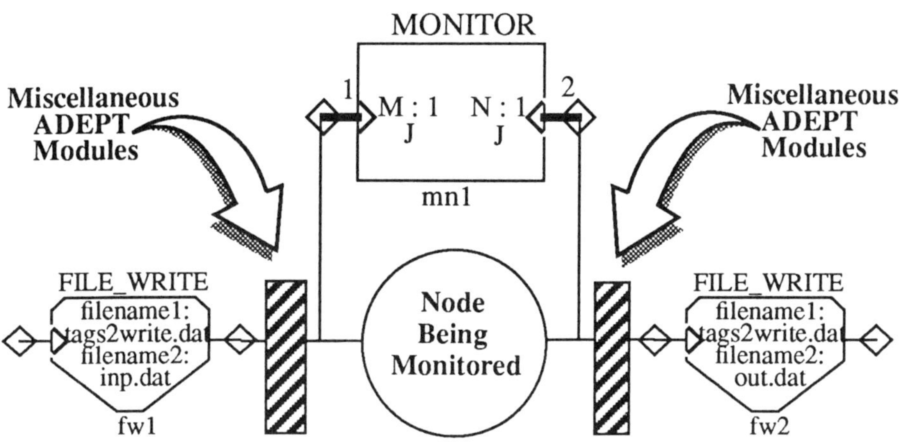

Figure 8.1 Use of the *MONITOR* module for bottleneck analysis

8.1.3 Performing Hardware/Software Trade-offs

The model can also be used to evaluate hardware/software trade-offs through the movement of functions from software into hardware and vice-versa. After performing such a trade-off, the conditions exhibited by equations (8.2)-(8.4) are true. Recall that the prime symbol indicates a new version of a description. The symbol V_{SM} represents the primitive operations of the software model, and V_{HM} corresponds to the operations supported by the hardware model. Equation (8.2) states that all of the operations in the software model are supported by the hardware model, which is true in general. Also, as indicated in equations (8.3) and (8.4), the new and the old virtual instruction sets for both the software model and the hardware model are different [263].

$$V'_{SM} \subseteq V'_{HM} \tag{8.2}$$

$$V'_{SM} \neq V_{SM} \tag{8.3}$$

$$V'_{HM} \neq V_{HM} \tag{8.4}$$

Hardware/software trade-off functions Γ can be utilized to express this movement of functionality from one domain to the other. The trade-off functions accept a software unit μ along with either a software function or a hardware function and produce a new software unit μ'. Given a software unit μ described using the abstract hardware/software model, the functionality (subprogram) corresponding to a software node n^S can be moved into hardware, resulting in a new software unit as shown in equation (8.5). As illustrated in Figure 8.2, the node and its constituent virtual instructions are replaced by a single virtual instruction in the software model, and a virtual instruction providing the functionality of n^S is added to a resource (execution unit) in the hardware model. Additional virtual instructions may be required for transferring operands to and extracting results from the resource.

$$\mu'_k = \Gamma(\mu_k, n^S) \tag{8.5}$$

As indicated in equations (8.6) and (8.7), new virtual instruction sets are produced for V_{SM} and V_{HM}. In V_{SM}, a new virtual instruction v_i corresponding to $n_i = n^S$ is added while some virtual instructions may be removed. The ones to be removed, indicated in the second part of equation (8.6), are the virtual instructions of n^S which are not utilized anywhere else by the function *f*. V_{swnode} corresponds to the virtual instruction set obtained through a decomposition of n^S. V_{pruned} refers to the virtual instruction set derived from *f* when n^S is removed. In V_{HM}, a new virtual instruction is only added, since, in general, existing virtual instructions in the hardware model need not be removed.

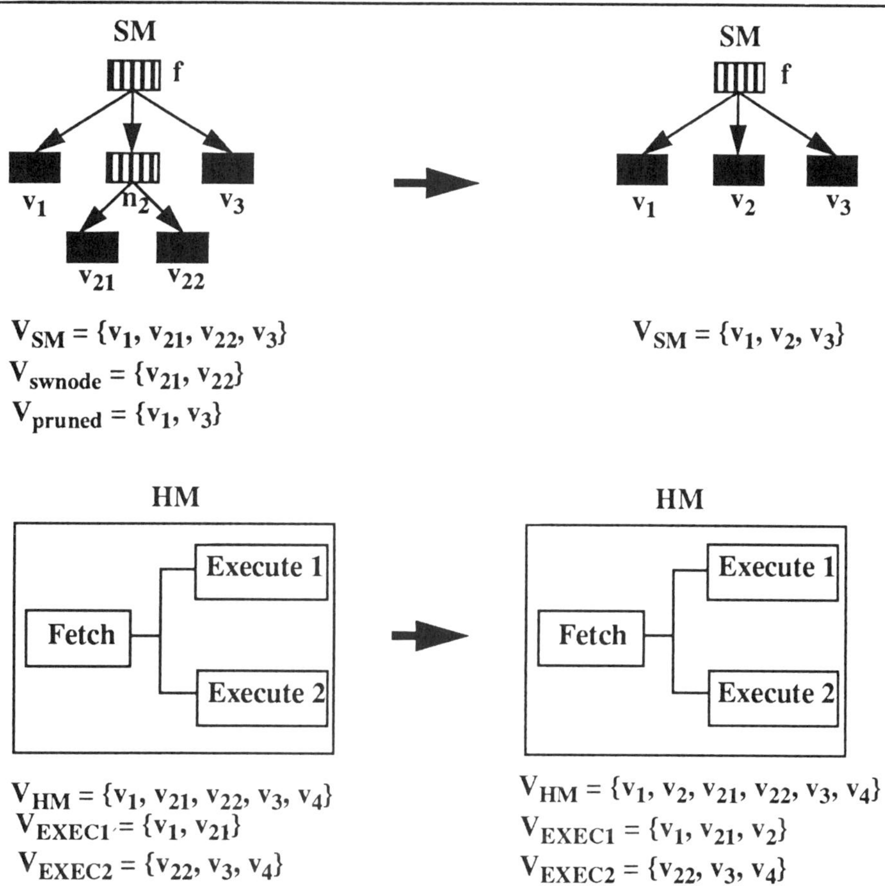

Figure 8.2 Example of a hardware/software trade-off

$$V'_{SM} = \left(V_{SM} \cup \{v_i\}\right) - \{v_j \mid v_j \in V_{swnode} \text{ and } v_j \notin V_{Pruned}\} \quad (8.6)$$

$$V'_{HM} = V_{HM} \cup \{v_i\} \quad (8.7)$$

In a similar manner, hardware functions can be moved into software as indicated in equations (8.8)-(8.10). A virtual instruction v_i in V_{HM} is expanded into a subprogram, consisting of primitive virtual instructions supported by the hardware model, whose topmost node is $n_i = n^h$. This subprogram replaces every virtual instruction v_i present in f. As a result, V_{SM} and V_{HM} are updated as shown in equations (8.9) and (8.10). In V_{SM}, any new virtual instructions associated with n^h (V_{hwnode}) are added, and the virtual instruction v_i is deleted. The same operations are performed with V_{HM}.

$$\mu'_k = \Gamma(\mu_k, n^h) \quad (8.8)$$

$$V'_{SM} = (V_{SM} \cup V_{hwnode}) - \{v_i\} \quad (8.9)$$

$$V'_{HM} = (V_{HM} \cup V_{hwnode}) - \{v_i\} \quad (8.10)$$

8.1.4 Evaluating Hardware/Software Alternatives

Given a function to be implemented, there are numerous alternatives that may need to be evaluated. These alternatives range from software executing on a general purpose processor, such as an Intel 80386, to software executing on a variety of application specific processors, for example, a Texas Instruments TMS320C30. At the extreme, a hardware implementation of the function may be employed. These alternatives vary in terms of performance, cost, reliability, and form factor, among others. The quantitative evaluation model can be used to evaluate the alternatives with respect to multiple metrics using a weighted technique. The abstract hardware/software model is used primarily for performance evaluation. However, it can also possibly be used to estimate other metrics as well.

8.1.5 Supporting an Integrated Modeling Substrate

Recall the integrated modeling substrate [194][195] depicted in Figure 8.3. Using a unified internal representation, the intent of this substrate is to support a wide range of hardware/software trade-offs, permit continuous and incremental hardware/software integration, allow more alternatives to be explored, improve model continuity, and support early evaluation. One possible application of the abstract hardware/software model is to serve as an integrated substrate during hardware/software development. As hardware and software are refined, evaluation can be performed at several stages of development, supporting many of the ideas discussed above.

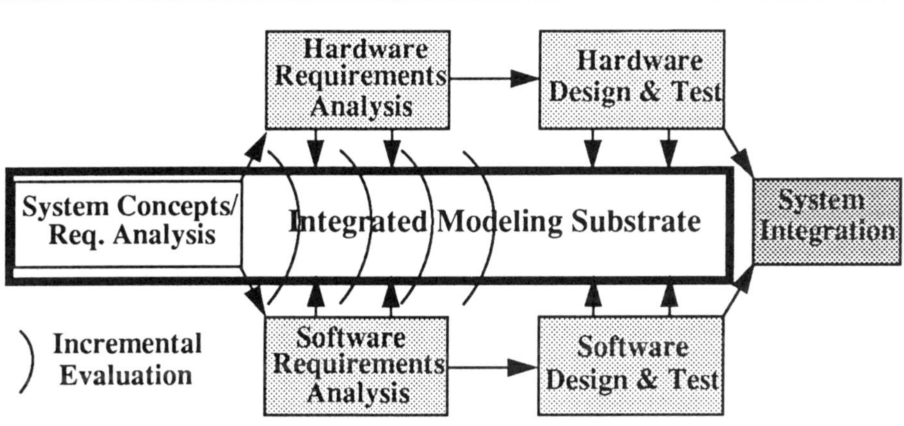

Figure 8.3 An integrated modeling substrate (Modified from [195], © 1992 IEEE)

8.2 EXAMPLES OF PERFORMANCE EVALUATION

Using several examples, this section shows how the abstract hardware/software model can be used to perform early evaluation and trade-off exploration. The examples are used to demonstrate many of the model's applications, which were discussed in the previous section.

8.2.1 Railway Control System

The example in this section illustrates several ideas regarding virtual instruction sets and specifications. A specification can simplify the design of the underlying implementation. Within a codesign context, a virtual instruction set derived through the decomposition of a software function can serve as a specification for the underlying hardware implementation. This specification can be used to develop simpler, application specific hardware. Note that the specification may include timing constraints as well.

Such a specification also serves as a contract between the software developer and the hardware developer. Thus, the underlying hardware implementation can be changed without "affecting" the software developer. Although this modification may impact software performance (or perhaps other nonfunctional attributes), the software should not have to be changed.

The example analyzes two different software units. In each software unit, a software model is executed on a processor model. However, one software unit utilizes a general purpose processor. The other employs an application specific processor.

System Overview. The V_Frame Architecture [264][265] is being designed to execute a control algorithm which consists of evaluating a series of Boolean expressions and accordingly setting switches and signal lamps within a railway system (see Figure 8.4). The railway system being considered is a simpler version than those found in actual systems. Nevertheless, it is representative. The system consists of three trains traveling on two track loops. The trains lock in routes from one platform to the other, and the control algorithm sets the control points (switches and signals) to ensure that no more than one train ever occupies a block of track.

The control algorithm consists of 8,496 equations whose operations can be divided into two classes: transfer operations and logical operations. Therefore, the algorithm does not contain any loops or branches. The transfer operations correspond to loads, stores, and moves. The logical operations consist of ANDs, ORs, and NOTs. These operations collectively define a virtual instruction set. Given this virtual instruction set, the control algorithm can be implemented in a variety of ways. One possibility is to execute the control algorithm on a general purpose processor. Another possibility is to utilize an application specific processor which supports only the operations within the virtual instruction set. At the extreme, one can develop an ASIC which implements the virtual instructions using logic gates.

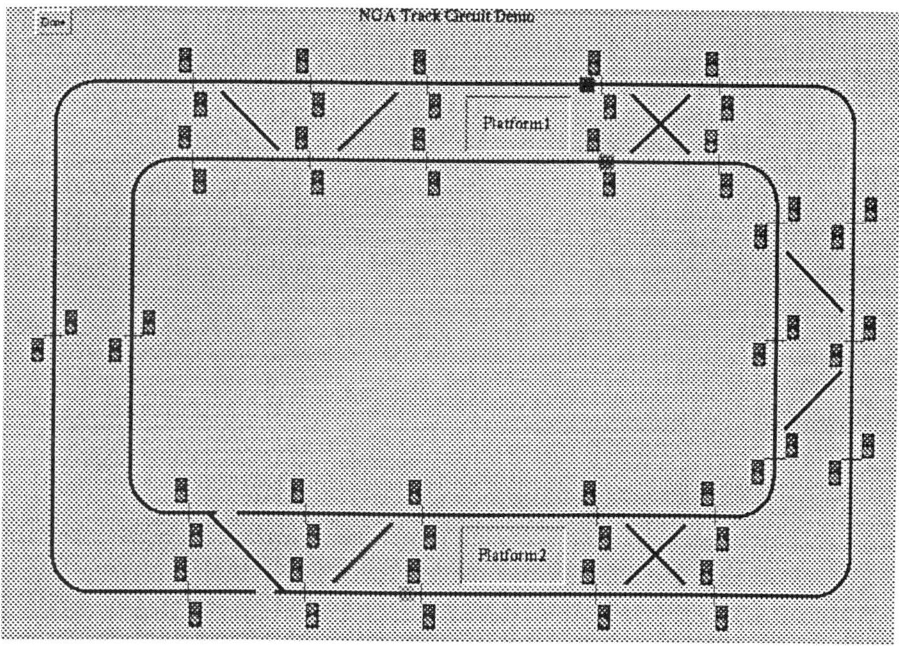

Figure 8.4 An example railway system [264]

Model Description. The software units were analyzed using the modeling structure shown in Figure 8.5. This particular model represents the control algorithm executing on an application specific processor. The *SOURCE* module (at left) outputs a token which initiates the abstract hardware/software model's (middle of figure) execution. The top rectangular symbol contains the software model, and the bottom one contains the hardware (processor) model. The parameters associated with the hardware model are shown below its corresponding symbol. In this example, three parameters are specified: the clock cycle time, the fetch delay, and the execute delay for an operation. Once the computation is complete, the abstract hardware/software model outputs a token to the *SINK* module (at right) which consumes the token.

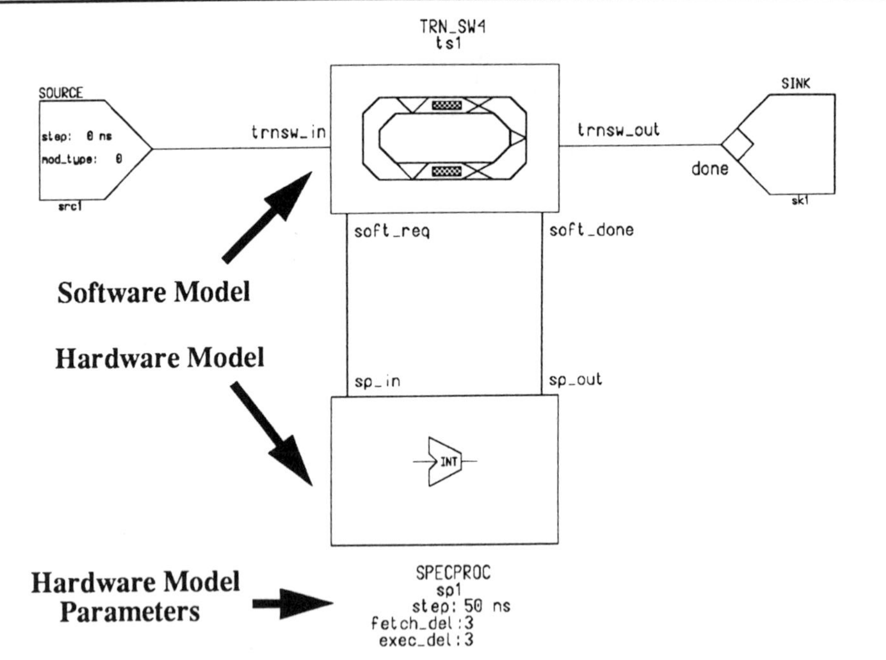

Figure 8.5 Top level view of abstract hardware/software model

As can be seen in Figure 8.6, a graphical representation is not used for the software model in this example. Instead, the software "requests" are stored in a file. Each request consists of the operation to be performed, a destination, and either one or two sources. The destination and source(s) correspond to the resources where the results are to be written to and read from, respectively.

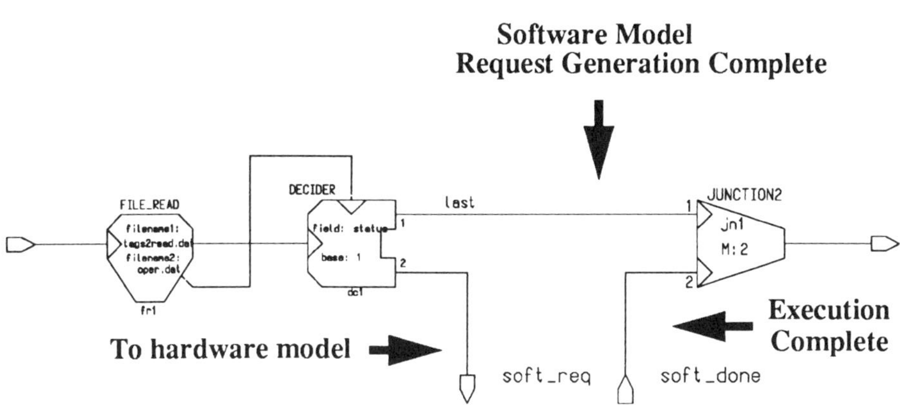

Figure 8.6 Modules used within the software model block

Upon the arrival of a token (from the *SOURCE* module) at its input, the *FILE_READ* module (at left) gets the first request from the file. This module populates the token with the information contained in the request and then sends this token to the *DECIDER* module (middle of figure). Using an output from the *FILE_READ* module, the *DECIDER* checks to see if an end-of-file condition has occurred. If such a condition has occurred, the token is routed to the *DECIDER's* top output. Otherwise, the request token is sent to the hardware model for execution via the bottom output.

The routing of the token to the *DECIDER's* top output, which corresponds to the top input of the *JUNCTION* module (at right), signifies that all of the requests have been generated and sent to the hardware model. The *JUNCTION* module requires that tokens be present on both inputs before a token is placed on the output. Once all of the requests have been executed, a token is placed on the bottom input of the *JUNCTION* module by the hardware model. This token is equivalent to a completion signal.

The hardware model is depicted in Figure 8.7. Requests enter a special element called the *SYNCHRONIZER* through the top input. This collection of ADEPT modules serves two purposes. First, the module passes the request token to the fetch stage (middle of figure). The fetch stage is exactly the same as the one described in Chapter 7. Once the software operation has been "performed" by the execute stage (top, right), a token is returned to the *SYNCHRONIZER* through a *BUFFER* module (bottom, right). The *BUFFER* module is used for controlling the token flow. At this point, the *SYNCHRONIZER* determines if all of the software requests have been executed by the hardware model. If all of the operations have been executed, a token is sent to the bottom output of the *SYNCHRONIZER*. This output is connected to the bottom input of the *JUNCTION* module in Figure 8.6. Otherwise, model execution continues as described above.

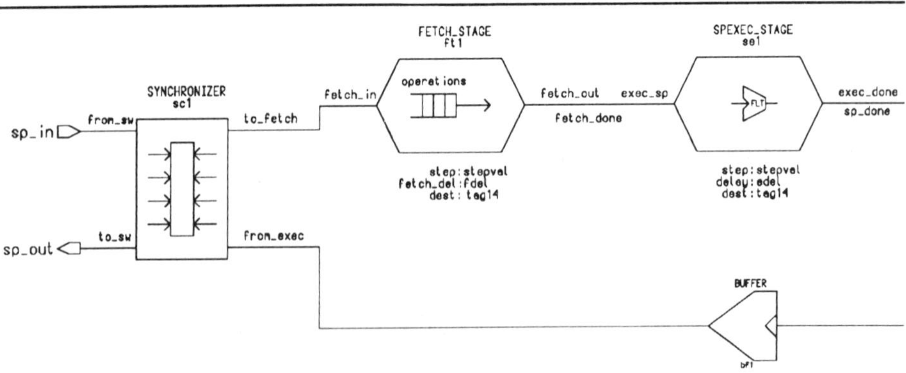

Figure 8.7 Hardware model

Referring to Figure 8.8, the execute stage of the hardware model contains a single hardware resource, the *ARITHMETIC_UNIT*, which is surrounded by two *BUFFER* modules to ensure proper token flow. The *ARITHMETIC_UNIT* is an uninterpreted, abstract resource which represents the computational delay associated with the virtual instructions. In this example, it is assumed that all of the virtual instructions take the same amount of time to execute. However, it is possible to specify the execution times individually. Note that parameters from the top level model (see Figure 8.5) are passed down to this element.

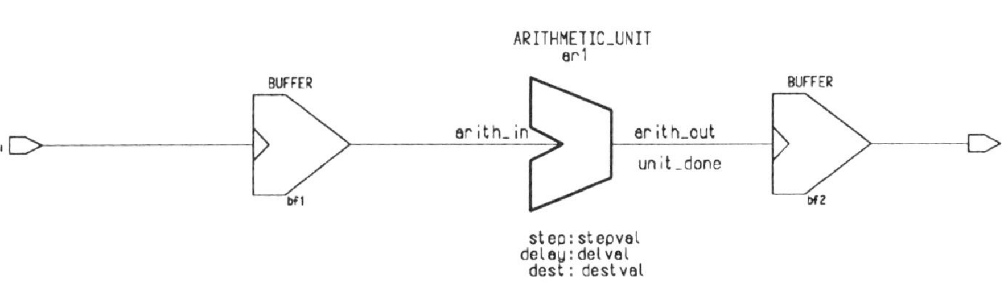

Figure 8.8 Execute stage of the hardware model

Simulation Results. Figure 8.9 displays the simulation results for the two software units. The first software unit used a general purpose processor (GPP). In this example, the execution times of the MC68020 microprocessor were utilized. It was assumed that the clock cycle time was 50 ns, the fetch time was 3 clock cycles, and all execution times were 6 clock cycles. In the second software unit, it was assumed that the execution time could be halved for an application specific processor (ASP) due to the simpler data path. Thus, in the application specific processor, the execution times for all operations were 3 clock cycles, with all of the other parameters remaining the same.

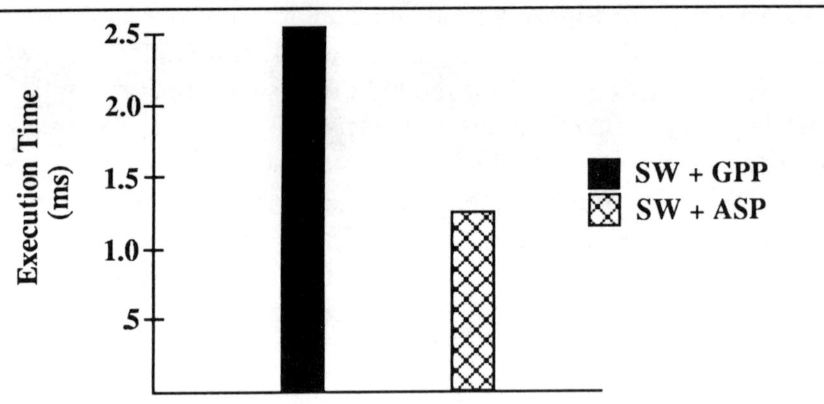

Figure 8.9 Simulation results for the software units

It is worth making a few observations about the software unit that contains the application specific processor. First, note that the application (control algorithm) has driven the development of the processor, producing simpler hardware which executes faster for the given application. However, although the application specific processor is functionally complete (AND, OR, and NOT are supported), it may be cumbersome to program larger, more complex control algorithms. If specifications are written precisely and followed rigorously, implementation changes should not require any changes to the software.

8.2.2 Aluminum Defect Detection and Classification

This example is used to illustrate the process of performing hardware/software trade-offs (see Figure 8.2). Starting with an abstract hardware/software model, bottlenecks within the software model are first identified using the concept of operator sensitivity discussed earlier in the chapter. Once these critical computations have been identified, functions can be moved from software into hardware, and the resulting abstract hardware/software model can then be evaluated. Thus, as indicated in equation (8.11), the example demonstrates the transformation of an existing software unit into a new software unit.

$$\mu_k' = \Gamma(\mu_k, n^s) \qquad (8.11)$$

System Overview. The example is a "best-fit ellipse" feature extraction algorithm used in a system for aluminum defect classification [266]. The major steps of the algorithm are shown in Figure 8.10. A defect image is obtained from an aluminum sheet using a camera and converted into a pixel representation. The orientation (angle θ) of the best-fit ellipse for the defect is determined, based on the object's center of mass and central moments. The orientation procedure performs one center of mass calculation, three central moment calculations, and one computation of θ (in that order). Using this information, the major and minor axes of the ellipse can be derived. These three pieces of information, the major axis, the minor axis, and the orientation angle, provide gross shape information which can be used for classification purposes.

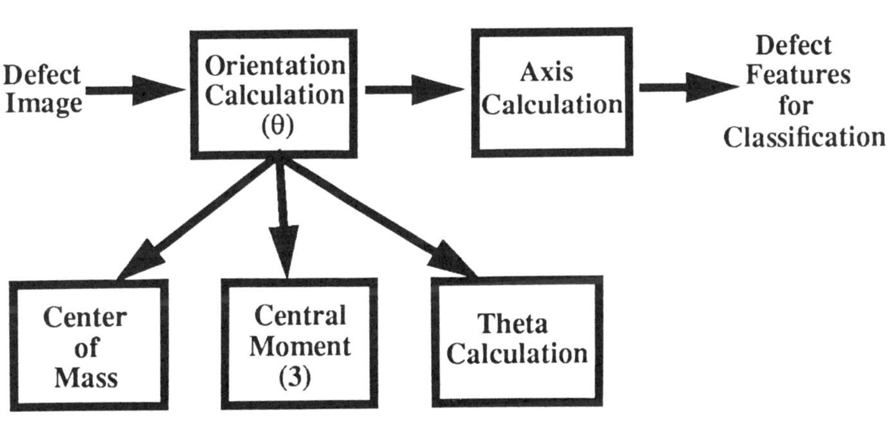

Figure 8.10 Best-fit ellipse feature extraction algorithm

Model Description. Figure 8.11 shows the top level model that was used to evaluate the best fit ellipse algorithm. Model execution is initiated by the *SOURCE* module (far left of model). This module simply outputs a token. A *SPLIT* module (right of *SOURCE* module), which is equivalent to the *WYE* module, sends this token along two different paths: to the *LOCK* module (top of model) and to the *FILE_WRITE* module (right of *SPLIT* module). In this example, the purpose of the *LOCK* is to enforce sequential execution of the abstract hardware/software model. In other words, only after the abstract hardware/software model has completed its execution will the *LOCK* allow the *SOURCE* to output a new token.

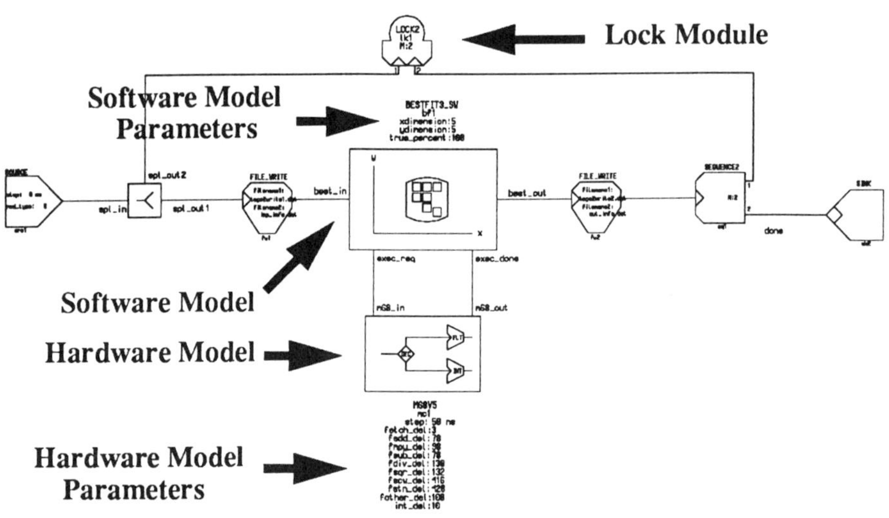

Figure 8.11 Top level of best fit algorithm executing on a processor

As described earlier, the *FILE_WRITE* module is used to aid in performing bottleneck analysis (see Figure 8.1). This module is the first of two such modules, writing the starting simulation time into a file. The abstract hardware/software model is found to the right of this

FILE_WRITE module, with the software model located above the hardware model. The second *FILE_WRITE* module, which writes the end simulation time to a file, is found to the right of the abstract hardware/software model. After the execution of the abstract hardware/software model is complete, a token is sent to the second *FILE_WRITE*. After writing the end simulation time to a file, this *FILE_WRITE* module sends a token to the *SEQUENCE* module. The *SEQUENCE* module allows the *LOCK* to enable a new execution of the abstract hardware/software model via the *SOURCE*. The *SEQUENCE* also produces a token which is consumed by the *SINK* module (far right of model).

Similar to the FIR example presented in the previous chapter, a control flow graph representation is used for the best fit ellipse algorithm. This software model is parameterized with the x and y dimensions of the defect image in pixels. The model is also parameterized with a "true" branch probability since a probabilistic branch node is utilized. This probability is used in a branch node which checks to see whether a pixel in the window determined by x and y is part of the defect image. In general, a defect image will only cover a portion of this window.

The orientation calculation of the best fit ellipse algorithm is shown in Figure 8.12. The top node corresponds to the center of mass calculation. The next three nodes represent central moment calculations. The execution of the central moment nodes is data dependent. All four of the nodes are hierarchical. Thus, the nodes are expressed in terms of more primitive nodes. The last set of nodes is used to compute the angle θ. These nodes are directly executed by the processor model.

A single routing element is found to the right of Figure 8.12. This routing element is a collection of ADEPT modules that "merges" the nine request lines, one from each of the nodes, into a single output line. This output leads to the processor model (perhaps through some other routing elements). When the model executes, a request token is generated from one of the nodes in the figure. This request is sent to the processor model via the routing element.

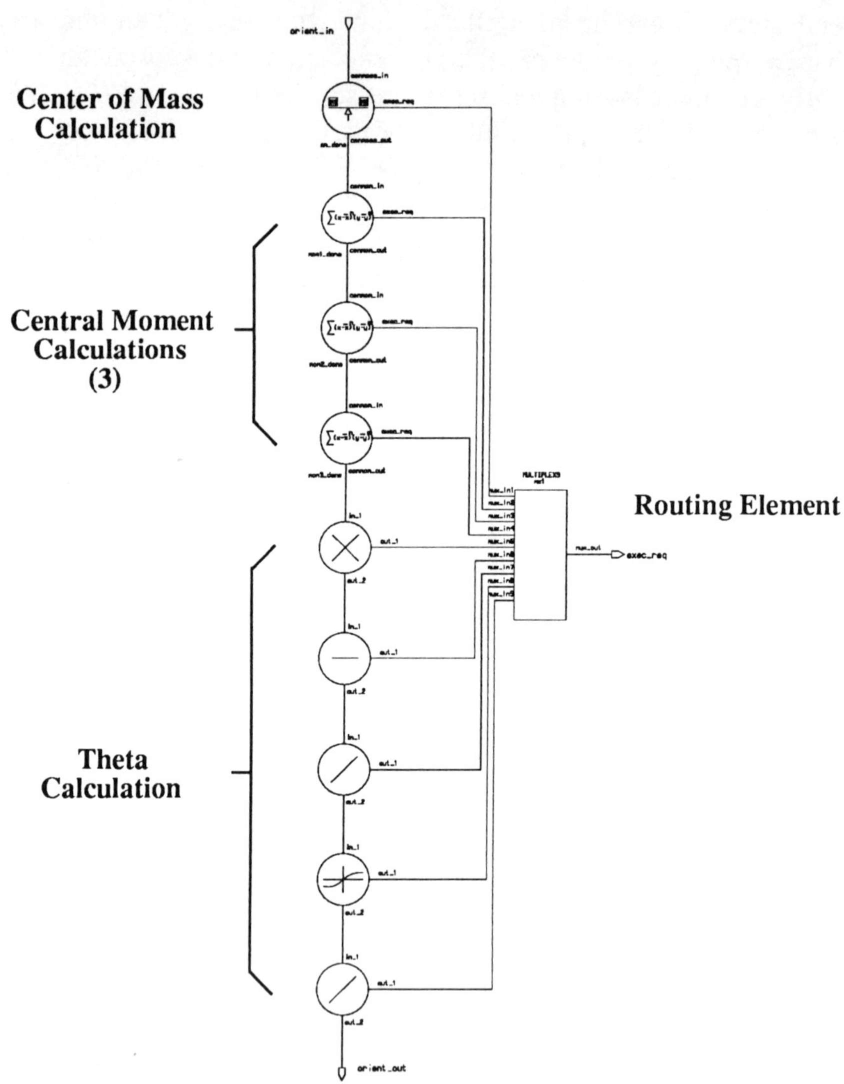

Figure 8.12 The orientation algorithm in terms of a control flow graph

Delay nodes can be used to improve the simulation time of an abstract hardware/software model through an "upannotation" process. In other words, a collection of nodes can be replaced by a single delay node, whose delay is equivalent to the sum of the delays of the nodes being replaced. For example, after determining the execution time of the orientation algorithm, a single delay node can replace the nodes in Figure 8.12.

The processor model is based on the MC68020/68881. The model contains a fetch stage, a decode stage, and an execute stage, which consists of two abstract execution units corresponding to resources that support integer and floating point operations. The decode stage (similar to a predicate node) determines whether the operation (request) should be executed by the integer unit or the floating point unit. The execute stage of the processor model is shown in Figure 8.13. Data book timings were utilized to parameterize the model.

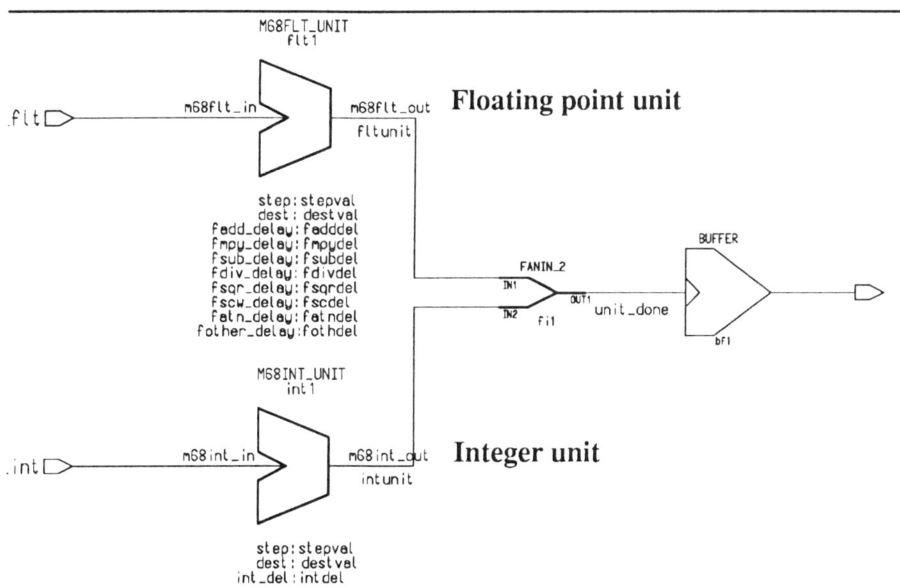

Figure 8.13 Execute stage of the 68020/68881 processor model

Simulation Results. Using the abstract hardware/software model, hardware and software engineers can collectively decide how best to speed up a computation. For example, faster resources may be incorporated. Alternatively, special purpose hardware can be employed. Of course, these benefits are achieved at some cost. Both of these approaches were explored in the example.

In the abstract hardware/software model, the software model was parameterized with the x and y dimensions set to 5. The true branch probability was set to 100%, reflecting a worst case execution time scenario. The model was described at different levels of detail, with some nodes represented using only delay elements.

Initially, only the orientation procedure was analyzed. It was determined through simulation that the operator sensitivity for a single central moment calculation was between 26%-34% of the overall orientation execution time, making this module a candidate for further improvement. A range of operator sensitivities resulted for the central moment calculation due to the data dependent execution of the module.

Figure 8.14 shows the percent improvement in the execution time of the orientation algorithm assuming a 20% improvement in the speed of various primitive operations: floating point multiply (FPM), floating point divide (FPD), and floating point arctangent (FPA). In many circumstances, an improvement in the execution of one operation can improve the execution of other operations if common resources are utilized. However, an assumption of this analysis is that the improvement of a single operation is independent of other operations. The disparity in improvement is a consequence of the operation frequency.

Another analysis was performed which included both the orientation calculation as well as the axis calculation. Referring to Figure 8.15, an operator sensitivity analysis revealed that the axis calculation consumed approximately 50% of the execution time. Using this information, a portion of the critical loop was moved into hardware and modeled as an abstract coprocessor.

It was assumed that this new coprocessor executed approximately twice as fast as the corresponding software. Also, no overhead was taken into account regarding the transfer of operands and results between the main processor and this new coprocessor, although this information can be incorporated into the model as well. The result of this movement produced an improvement of approximately 23% in the execution time of the algorithm. However, communication overhead will diminish this improvement.

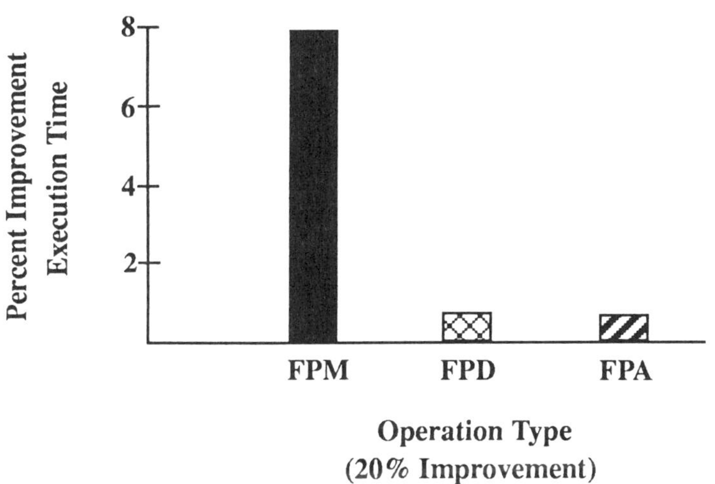

Figure 8.14 Percent improvement in orientation algorithm

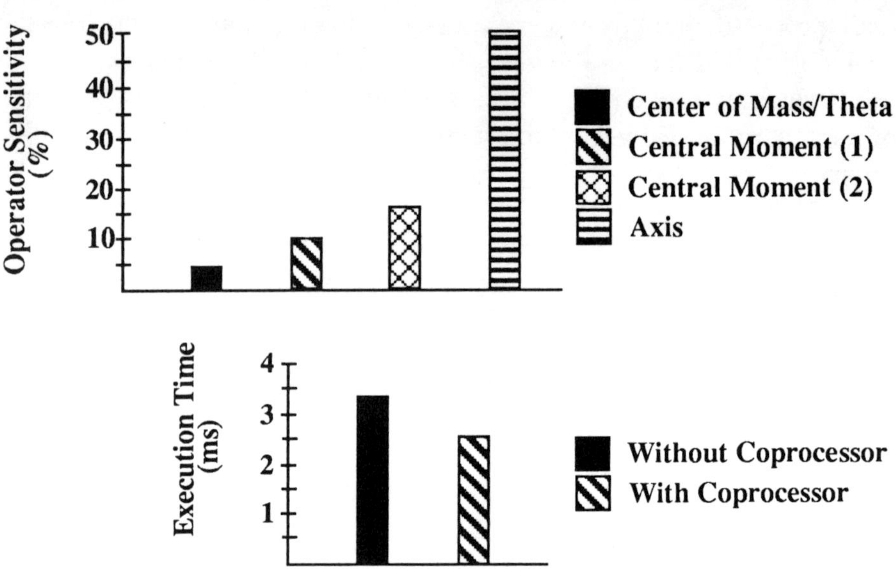

Figure 8.15 Hardware/software trade-off analysis

8.2.3 A Stylus Tracking System

In this section, the abstract hardware/software model is used to assess the performance of several hardware/software alternatives for a function. Unlike previous examples, these hardware/software descriptions are evaluated within the context of a system model which uses timing constraints. One point of this example is that abstract descriptions can be utilized for certain portions of the system while more detailed descriptions can be employed for those portions of interest. Also, this example demonstrates certain aspects of the codesign methodology presented in Chapter 5.

System Overview. A system [267] is being developed to track light movement across a screen. The system is an example of a real-time application whose major processing steps are illustrated in Figure 8.16. A position detector outputs four analog (electrical) current values, up (**U**), down (**D**), left (**L**), and right (**R**), based upon the position of a light emitter on a screen. These four current "directions" are sampled N_S times at a rate of f_S, corresponding to a sampling interval of T_S. After performing analog to digital (A/D) conversion of the sampled data, the values are transferred to memory. Next, a spectral analysis is performed on these values using a Fast Fourier Transform (FFT), once for each direction. The amplitudes of the FFT data are used to update the **(x,y)** position of the light on the screen. A constraint imposed on the system is that the position update, that is, the entire process shown in Figure 8.16, must be performed every 1/30 of a second, or 33 milliseconds (msec).

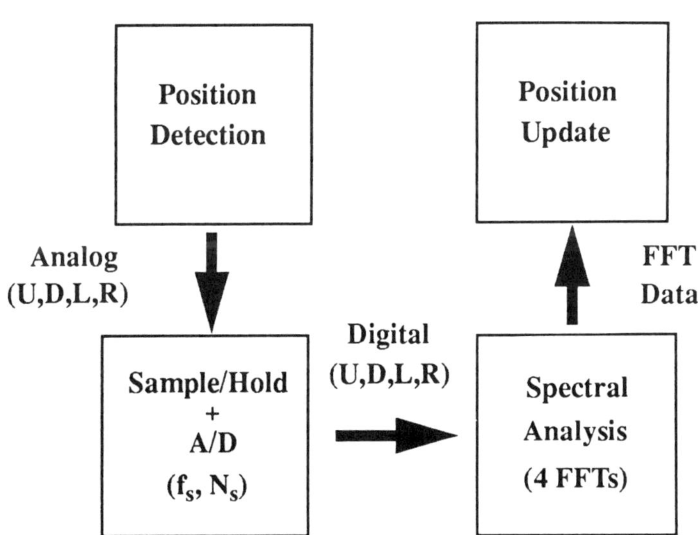

Figure 8.16 Processing steps in stylus tracking system

The most time consuming steps within the system are the sampling and the FFT computation. Of these two computations, the FFT is the most intensive. The remaining portions of the system take a relatively small amount of time and, as a result, can be neglected in the performance analysis. Given f_S and N_S, a timing constraint can be established for the FFT. Assuming that f_S=100kHz and N_S=1024, the time to perform the FFT computation, T_{FFT}, is approximately 20 msec.

Model Description. A model of the system was constructed to analyze several alternatives for the FFT. A top level description of the model is shown in Figure 8.17. In a manner similar to the previous model, a *SOURCE* module (left of model) is used to initiate execution, and the *LOCK* module (top of model) ensures sequential execution. The two parameterized symbols, third and fourth from the left, correspond to the sample and hold/analog to digital conversion and the FFT function, respectively. A block consisting of ADEPT modules is used to check if all four FFTs have been completed. To aid in the analysis, a new ADEPT module called the *TIMECHKR* (second from top), which can be parameterized with a user-specified timing constraint, was developed to determine if any constraints were being violated for the FFT. The usage of this module is similar to that of the *MONITOR*. If a timing violation is detected, the *TIMECHKR* places a token on its control output. The control output of the *RC* (read color) module, shown as a dotted line in the figure, is used to reset the *TIMECHKR* if any violations have occurred.

The system functions being considered can be partitioned into hardware and software. The sample and hold/analog to digital conversion function can be performed by a hardware unit, which accepts data as input and produces data as output. The FFT function can be implemented in software, using one or more software units.

Referring to Figure 8.18, the sample and hold/analog to digital conversion block is modeled using ADEPT modules. The model is parameterized with a sampling interval T_S and the number of samples to be taken N_S. This abstract model captures the essential performance characteristics of the process, which is performed by hardware.

However, in this ADEPT description, only the high level, abstract behavior is reflected.

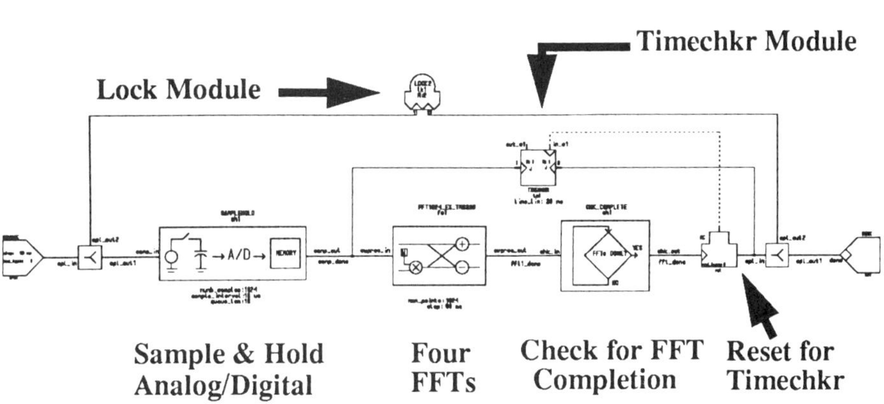

Figure 8.17 Top level of model for the stylus tracking system

The high level behavior of this model is now described. A token, representing the four analog currents to be sampled, arrives from the left of the figure. This token is sent to four abstract sample and hold elements (four parallel rows in the middle of the figure). Each element performs the sampling for a particular current direction (up, down, left, and right).

Note that this sampling occurs in parallel, implying that four sample and hold circuits exist. These four sample and hold elements along with the collection of ADEPT modules at the bottom of the figure model the process of sampling the analog signal, performing the analog to digital conversion, and interrupting the main processor. Delays are included in each element to account for the time required to perform these activities.

The interrupt process is modeled using the *C_AND* (control AND) module. After all four circuits have sampled their signal, an "interrupt" is generated. In the actual system, the processor stores these values into memory at this point. In this model, no explicit transfer of data by the processor is shown. After all 1024 samples have been generated, each sample and hold element sends a token to the *QUEUE* module (far right of figure).

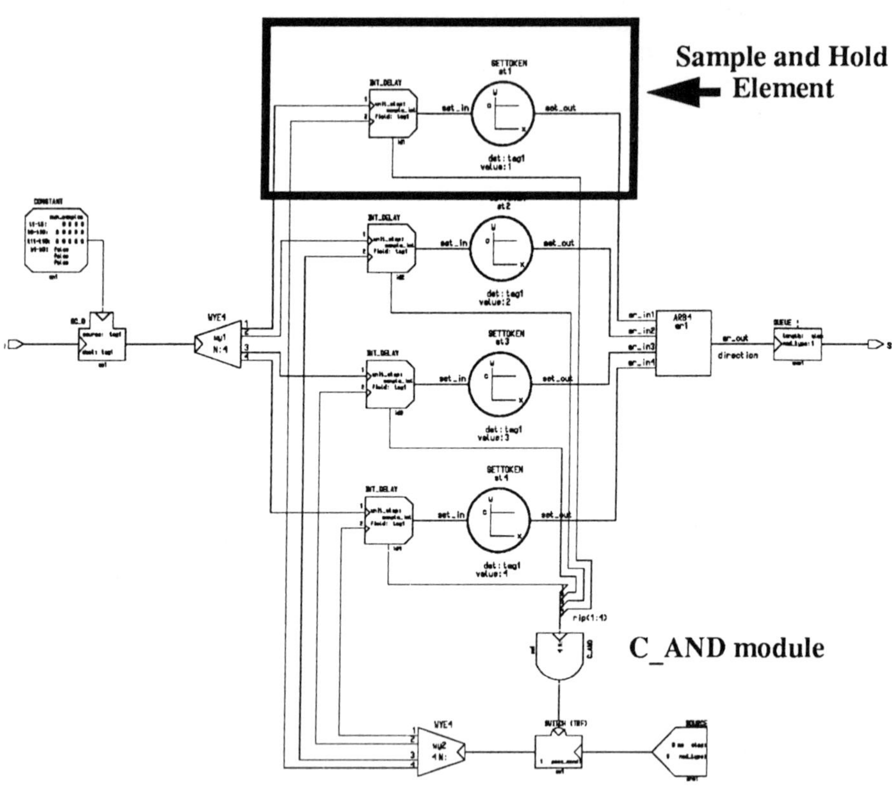

Figure 8.18 Sample and hold/analog to digital conversion block

The FFT block contains the abstract hardware/software model. Four FFTs are performed, one for each token generated by the sample and hold process (that is, for each token stored in the *QUEUE* module). This block is parameterized with the number of points N associated with the FFT, along with the parameters of the specific processor model. The software model consists of two loops, one nested within the other. The outer loop is executed log_2N times, while the inner loop is executed $N/2$ times. The inner loop consists of the operations found within a butterfly computation.

Simulation Results. There are several possible ways of implementing the FFT function. These alternatives span a spectrum of different mixtures of hardware and software from general purpose to application specific solutions. As an example, an abstract hardware/software model of a decimation-in-time FFT algorithm executing on a general purpose processor was developed. The processor selected was a 20 Mhz MC68020/68881. A fetch delay of three clock cycles was assumed. In this model, no concurrency was exploited since the software model consisted of only floating point operations within a butterfly computation, which could only be executed serially with the MC68881 floating point unit. Through simulation, an approximate analysis revealed that a 1024 point FFT would take 220 ms, violating the time constraint.

The performance of the FFT can be improved in several ways. One way is to use a specialized processor, such as a digital signal processor. Because the FFT consists of several butterfly computations, another possibility is to utilize an even more specialized butterfly processor [268]. Yet another alternative is the use of multiple processors. At the extreme, a “hardware” (with perhaps some microcode) implementation of the FFT can be developed. However, a hardware implementation of a 1024 point FFT can potentially be costly, not to mention the overhead associated with getting operands in and results out. The simulation results for several of these different hardware/software alternatives are provided in Figure 8.19.

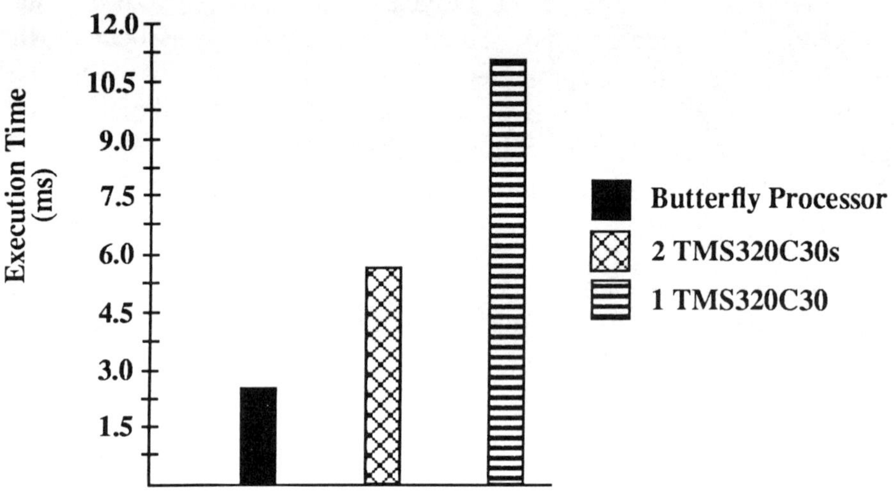

Figure 8.19 Simulation results for FFT alternatives

The TMS320C30 processor models used a single-cycle execution time of 60 nsec, whereas the butterfly processor employed a 15 nsec single-cycle execution time. For the two processor case, it was assumed that the data to perform the FFTs was available locally. The execution times represent the time to perform four, 1024 point FFTs, where each FFT consists of only butterfly computations. In the two processor alternative, each processor performed two FFTs. For these alternatives, a more detailed description of the FFT is required to assess violations of real time constraints.

The two processor alternative is shown in Figure 8.20. The tokens representing the up/down samples are extracted from the *QUEUE* in Figure 8.18 and sent to the top abstract hardware/software model. Similarly, the left/right tokens are sent to the bottom model.

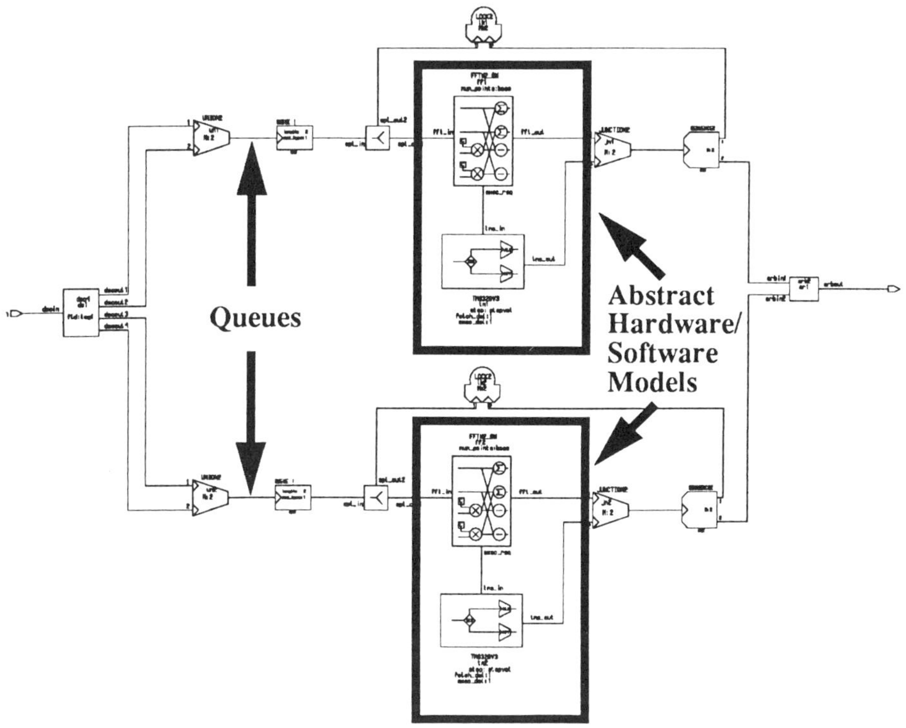

Figure 8.20 Two processor alternative for the FFT function

8.2.4 A Distributed System for Parallel Discrete Event Simulation

This section presents the model and the simulation results for a distributed system used to support parallel discrete event simulation. In this example, functional descriptions, independent of hardware or software, are used in the analysis. Also, the different steps of the codesign methodology are related to the example. Another type of hardware/software trade-off is presented as well.

System Overview. *Parallel discrete event simulation* (PDES) [269], also called distributed simulation, refers to the execution of a discrete event simulation program on a parallel machine. In a PDES, a simulation program is decomposed into a collection of concurrently executing processes that communicate and synchronize. In a discrete event simulation model, the system being simulated changes state at discrete points in time due to the arrival of an event. For example, in the simulation of a queuing system, the state of the queues within the network is affected by the arrival and departure of tokens. The arrival and departure of tokens are events that correspond to the arrival of a token into a queue and the departure of a token from a server, respectively, and produce changes in the lengths of various queues.

Referring to Figure 8.21, a hardware/software framework [270] is currently under development at the University of Virginia to support parallel discrete event simulation. The PDES framework consists of host processors (HPs), such as SUN SPARC workstations, which communicate via a host communication network (HCN) using messages. The HPs execute a discrete event simulation algorithm and interface to auxiliary processors (APs) using a dual-ported RAM (DPRAM). The APs execute synchronization algorithms and exchange information with the aid of a parallel reduction network (PRN) through a register interface (IN/OUT). The PRN is a synchronization network, consisting of a binary tree of pipelined ALUs, which can rapidly compute and disseminate information to the HPs. The ALUs are programmed to perform binary, associative operations, such as sum, minimum, maximum, logical AND, and logical OR.

The PDES framework incorporates an interesting hardware/ software trade-off. The ability to quickly compute and disseminate global synchronization information is important in reducing the total time required to perform a PDES. One option is to perform global operations, such as minimum or maximum, using software running on the HPs. However, the time to perform global reductions on existing parallel architectures can be costly. For example, the time to perform a global reduction operation using barriers for a 32 processor Intel iPSC/2 is on the order of 10 milliseconds [271]. Another approach, represented by the PDES framework, is to perform the global operations in special purpose hardware (the PRN).

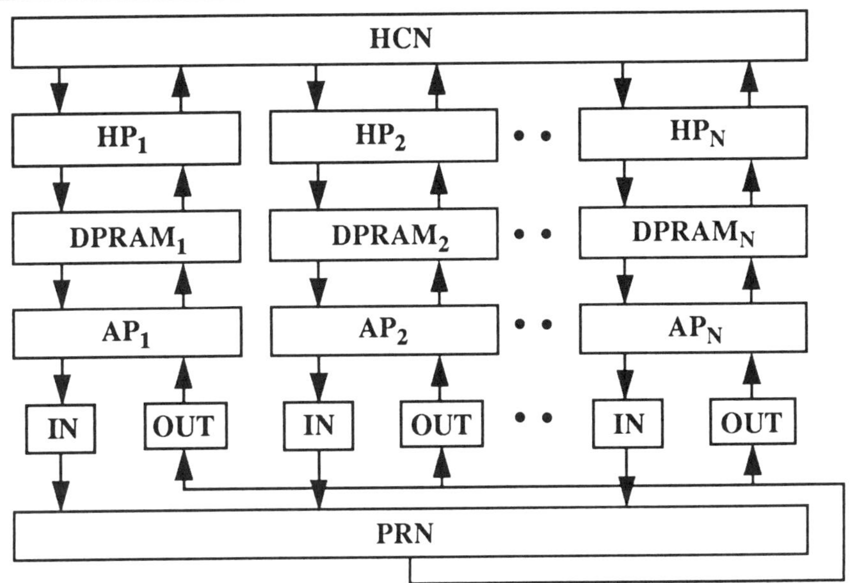

Figure 8.21 Hardware/software framework for PDES (From [270],© 1994. Reprinted by permission of Academic Press, Inc.)

Model Description. The codesign methodology can be used to design the framework. Initially, the framework can be viewed as a collection of system functions, independent of hardware or software, which communicate and synchronize. In the terminology of the codesign methodology, this description corresponds to the system representation. System partitioning maps the system functions onto physical units. After performing codesign, the framework can be considered an implementation for the abstract function PDES, consisting of several software units and a single hardware unit. The HPs and APs are software units, and the PRN, a free running machine that continuously accepts input data and produces output data, is a hardware unit.

An eight node (N=8 HP/AP pairs) shared bus model for this framework was developed. Except for the HPs and the APs, ADEPT modules were used to model all portions of the framework. The full simulation model, as seen in the ADEPT environment, is shown in Figure 8.22. The top portion of the figure corresponds to the HCN. A delay of 10 microseconds was utilized to model the communications time between the HPs. A collection of queues exists under the HCN. The next four rows of the model correspond to the HPs, the DPRAMS, the APs, and the IN/OUT registers, respectively. The bottom portion of the model displays the PRN and associated ADEPT modules used to distribute the PRN outputs to the APs via the OUT registers of the register interface.

The HPs and APs were described as concurrent processes written in custom VHDL. Delays were inserted within these descriptions to model the time to perform various computations. Also, code was included that allowed them to be interfaced with other ADEPT modules.

Generics were used to initialize the HPs with a list of "events". The HP algorithm consisted of a loop which obtained the next event to be processed, "executed" the event, and sent a message to another HP. The execution delay of an event was estimated to be 100 microseconds. Messages were sent to the APs when the HP's logical (local) clock was updated, a message was sent to another HP due to the execution of an event, or a message was received from another HP through the HCN. At the end of the loop, the HP algorithm checked to see if any data was received from the auxiliary processor. With the aid of the PRN, the AP algorithm performed acknowledgments of messages sent by the HP.

The PRN model (see Figure 8.23) was simplified, supporting only the minimum operation. The model consists of three pipelined stages (levels) since N=8. Each ALU within the PRN performs a minimum operation on its inputs and has a stage delay that can be parameterized.

In the model, this delay was set to 150 nanoseconds. Note that these ALUs perform functional transformations on the inputs, and therefore, are not the same as the abstract resources used in earlier examples.

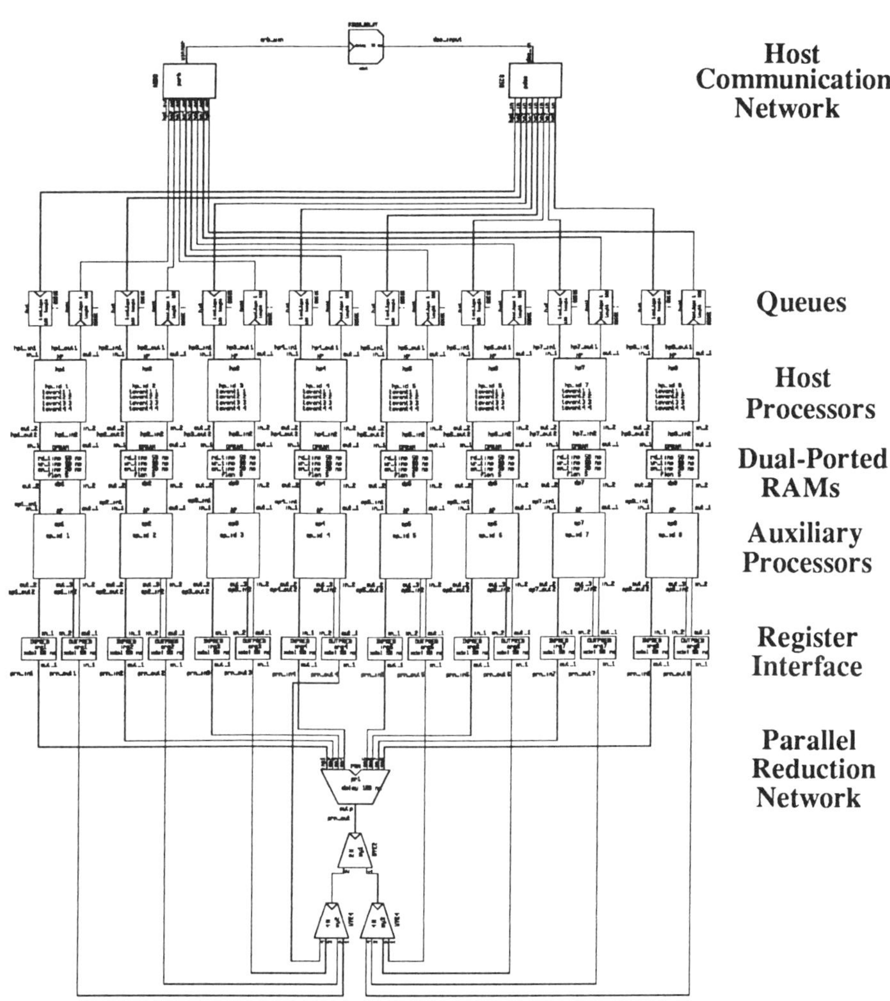

Figure 8.22 PDES framework simulation model

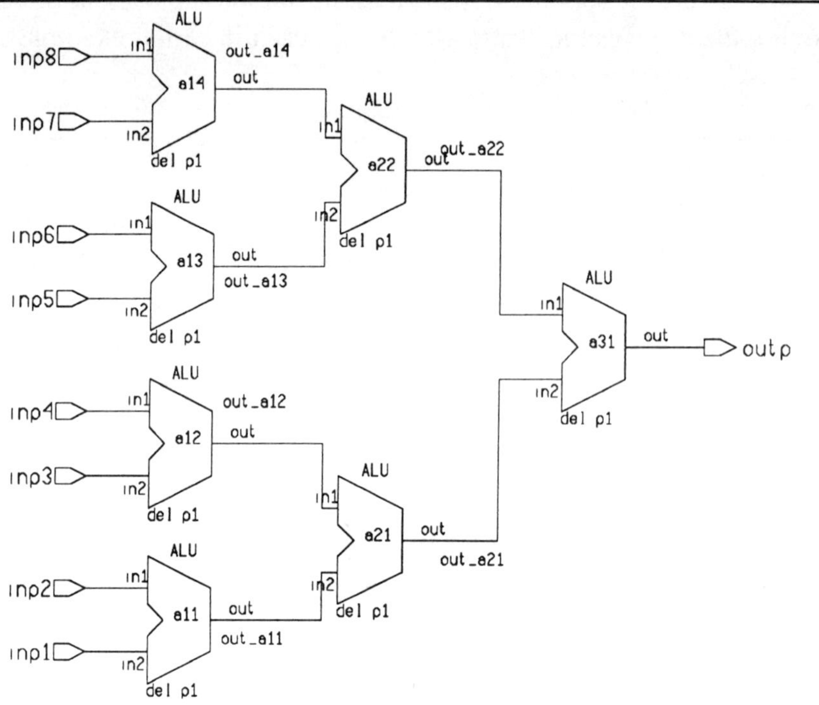

Figure 8.23 Parallel reduction network

Simulation Results. To reduce message traffic in the HCN, the APs and the PRN are utilized for performing acknowledgments of messages sent between HPs. A two phase acknowledgment protocol, requiring two global reductions through the PRN, is used for message acknowledgments. When more than one AP tries to perform acknowledgments simultaneously, message acknowledgments become serialized through the PRN. In the worst case, all APs perform acknowledgments simultaneously. Thus, an important issue is analyzing the effect of this serialization on the performance of the framework.

The model was used to analyze the effect of serialization on the framework. For the worst case scenario, the simulation results in Figure

8.24 display the serialized acknowledgment times for messages sent between HPs in "chained" fashion (1 to 2, 2 to 3, and so on). In other words, after executing an event, an HP would send a message to its neighbor on the right.

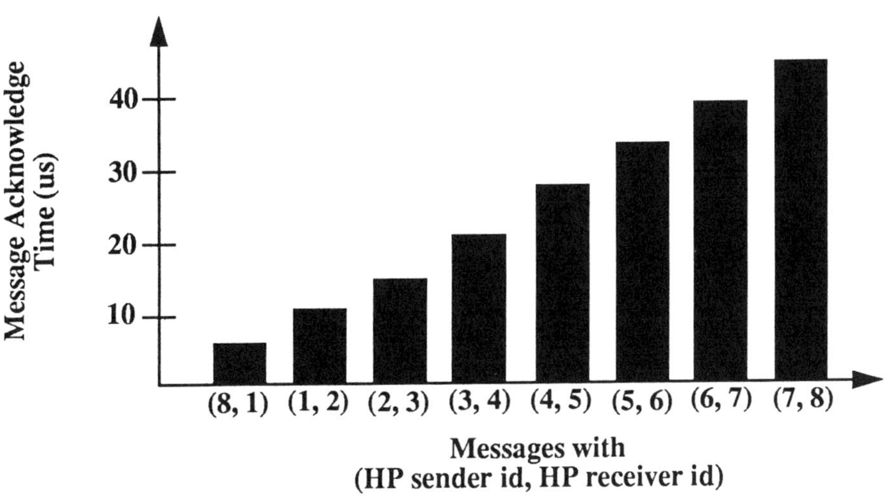

Figure 8.24 Serialized message acknowledgment times

8.3 SUMMARY

This chapter started by describing the applications of the abstract hardware/software model. The use of operator sensitivities to identify bottlenecks and the process of performing hardware/software trade-offs were discussed. Using a collection of examples, the next section demonstrated many of these applications. Specifically, it was shown that the model could be used to identify software bottlenecks, evaluate hardware/software trade-offs, and evaluate design alternatives.

The examples in this chapter have demonstrated several ideas regarding virtual instruction sets. The virtual instruction set can be used as a specification, serving as a contract between hardware and software developers. Also, the process of performing hardware/software trade-offs can be easily understood using this concept. The application specific approach to processor design shares much in common with hardware synthesis capabilities since both can be described in terms of virtual instructions.

Chapter 9
Object-Oriented Techniques in Hardware Design

Chapter 7 and Chapter 8 illustrated a unified representation based on functional abstractions. This chapter demonstrates that data abstractions, which form the basis for object-oriented techniques, also provide a uniform way of representing both hardware and software [250]. A unified representation for hardware and software allows techniques from one domain to be applied to the other domain. Therefore, by using a unified representation based on data abstractions, object-oriented concepts, used primarily in the software domain, can be employed in the hardware domain as well. Also, existing software techniques, such as those used to verify the correctness of abstract data type implementations, can be utilized for hardware.

The examples presented in this chapter are written in C++ [272], which supports user-defined data types and inheritance. The reason for utilizing C++ is to demonstrate the benefits of object-oriented techniques, not to provide arguments for or against the use of the language for hardware modeling.

9.1 MOTIVATIONS FOR OBJECT-ORIENTED TECHNIQUES

The application of object-oriented techniques to hardware design is an example of cross fertilization from software engineering to hardware engineering. There are several motivations for exploring these techniques. With the advent of hardware description languages and the increased use of simulation and modeling in hardware design, it makes sense to look at programming techniques that can improve the modeling process. In many circumstances, individuals that model hardware take on the role of software programmers. Object-oriented techniques can also aid in the management of hardware complexity. Hardware complexity management and change management issues are becoming increasingly important [181].

Although object-oriented techniques may seem foreign to hardware designers, many of the underlying concepts and principles are familiar. In particular, it is natural to think of hardware resources as components which consist of state and a collection of associated operations that can manipulate this state. In addition, the construction of systems using reusable library components, a highly touted advantage of object-oriented techniques, manifests itself in the hardware domain as design using off-the-shelf building blocks [273][274]. The use of hardware building blocks allows systems to be constructed quickly, lowers overall design cost, and increases reliability.

The advantages of object-oriented techniques when applied to hardware design include:

- improved modifiability and maintainability of models;
- easy component instantiation with different parameters;
- tailoring of general purpose components to more specialized components;
- quick composition of new components;

- the ability to identify and reuse common components;
- support of dynamic object creation and destruction; and
- the possibility of employing existing software synthesis and verification techniques.

All of these benefits derive from the use of data abstractions as a unified representation for hardware and software.

9.2 Data Types

Before illustrating the application of object-oriented techniques to hardware design, it is worthwhile to briefly discuss some fundamental ideas behind data types. Recall that a data type consists of a domain of values and a set of operations. An operational specification (or abstract model) is used to define data types. Such a specification for a data type has two parts: a domain specification, which is an abstract description of the type, and a specification of the abstract operations that can be performed on objects of that type.

In addition to function, data types are present in both software and hardware. Programming languages generally provide a set of built-in data types, such as integer, character, boolean, and real. In the same way, machines also support data types. Bit, integer, and float are certainly common, although machines which support vector computations and string manipulations also exist.

Some programming languages support the creation of *abstract data types* (data abstractions), user-defined data types that represent some abstraction of a real life entity. In the same manner that new functions, such as square root, can be added to extend the virtual machine defined by a programming language, new data types, for example, queues and lists, can be added to the virtual machine as well [117]. This point will become more clear later.

9.3 MODELING HARDWARE COMPONENTS AS CLASSES

Data abstraction can be used to represent hardware. C++ supports data abstraction through the concept of a *class*. In general terms, a class corresponds to a set of elements with common characteristics. Thus, a hardware component can be treated as a class containing state along with a collection of associated operations that can manipulate this state. For example, a register can be viewed as a class with the operations read and write. The contents of a register correspond to its state, which can be accessed and manipulated using the operations read and write, respectively. Although combinational elements do not possess state, the concept of a class can still be used to capture the common properties of these devices. Thus, an arithmetic logic unit (ALU) class may define the operations add, subtract, logical AND, and shift. At a higher level of abstraction, a processor contains state, consisting of the values of the program counter and other internal registers, which is manipulated by its supported instructions. These three hardware classes are shown in Figure 9.1.

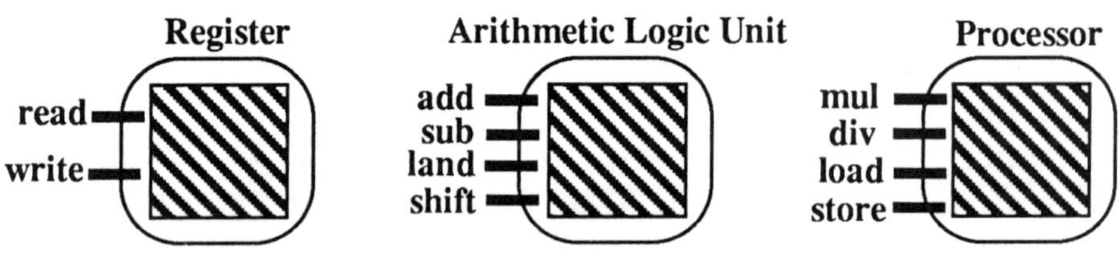

Figure 9.1 Examples of hardware classes

A C++ description of the class register (class reg) is shown in Figure 9.2. An abstract description of a register is contained in the region labeled "protected". The label "protected" allows classes which are derived from class reg to access this information directly. Other classes must utilize the member functions (operations) found under the label "public", which indicates that the functions are accessible by anyone.

```
class reg {                              // class register
protected:
        int contents;                    // integer contents of register
        int num_bits;                    // number of bits in representation
        BitString bitrep;                // bit representation of contents
public:
        reg(int size = 16)               // default size of 16 bits
        {
            contents = 0;                // initialize contents to zero
            resize(size);                // create a particular size register
        }
        void resize(int size)
        {
            bitrep.clear(0, size-1);     // clear all of the bits
            num_bits = bitrep.length();  // obtain number of register bits
        }
        int READ()                       // read the register contents
        {
            return(contents);
        }
        int WRITE(int newval)            // write a value into the register
        {
            contents = newval;           // update the contents
            write_regbits(newval);       // write the bit representation
        }
}
```

Figure 9.2 C++ description of a register class

A stronger level of data hiding is provided by treating the abstract description as "private". By considering the data as private, even derived classes cannot access the abstract description directly. Regardless, the abstract description and the member functions collectively constitute what is common to all registers.

There are three pieces of "protected" information contained in this class: contents, num_bits, and bitrep, which indicate the integer contents of the register, the number of bits associated with the register, and the bit representation of the register, respectively. The data item bitrep is an object of class BitString [275] whose instantiation (creation) is performed by the third line under the "protected" label. More precisely, bitrep is instantiated only when a register object is instantiated.

The most important member functions are READ() and WRITE(). These member functions allow the protected information to be read or written, respectively. The reg() member function is called a constructor and is executed when a register object is instantiated. By convention, a constructor has the same name as the class.

In this constructor, a default size of 16 bits is assumed if no size is specified for an instance. When an object of class reg is instantiated, the constructor initializes the integer contents of the object to zero and calls the member function resize(). Resize() initializes the bit representation to all zeros and sets the number of bits for the object using the member function invocations bitrep.clear() and bitrep.length(), respectively.

The description in Figure 9.2 is a specification of a register. Nothing is said about how the register is implemented. For example, the implementation of the register would consist of some collection of gates and flip-flops.

An object of class reg must be created before being used. As demonstrated with bitrep, object instantiation occurs in the same manner as variable declaration. In Figure 9.3, three register objects are created within the main() program: the condition code register (CCR), the memory address register (MAR), and the memory data register (MDR).

The size of the registers in bits is specified in parentheses. Thus, the same code can be reused to create different sizes of registers with the same basic properties. Once the objects are created, various operations can be performed, such as the reading of the memory address register (see Figure 9.3).

```
main()
{
        reg ccr(3);                          // create a 3-bit CCR
        reg mar(16);                         // create a 16-bit MAR
        reg mdr(16);                         // create a 16-bit MDR

        int contents = mar.READ();           // read contents of MAR
}
```

Figure 9.3 Creating register objects through instantiation

Although not shown in Figure 9.2, it may be desirable to parameterize the model with the physical attributes of the register. Some possible attributes may include read/write delays, area, and power consumption. If necessary, this information could be extracted using additional member functions, such as read_delay(), write_delay(), area(), and power(). Note that a kind of software abstraction, namely data abstraction, has been used to describe a hardware abstraction at the register-transfer level. It is also possible to use data abstraction to describe hardware at other levels of detail, such as the logic level.

9.4 DERIVING SPECIALIZED COMPONENTS

Starting with a collection of base classes, it is possible to derive more specialized classes of components through inheritance. As shown in Figure 9.4, the register class can be used to derive more specialized registers. Specifically, a program counter is a register with an increment (incr) operation, and a stack pointer is a register with an increment (incr) and decrement (decr) operation. An instruction register can be viewed as a special register in which the contents are divided into various fields which represent the opcode, the source operands, and the destination of the result. Additional member functions would be required to extract these individual pieces of information from the instruction register. These newly derived classes obtain the abstract description and member functions of class reg. Thus, all of these registers have an integer contents and a corresponding bit representation.

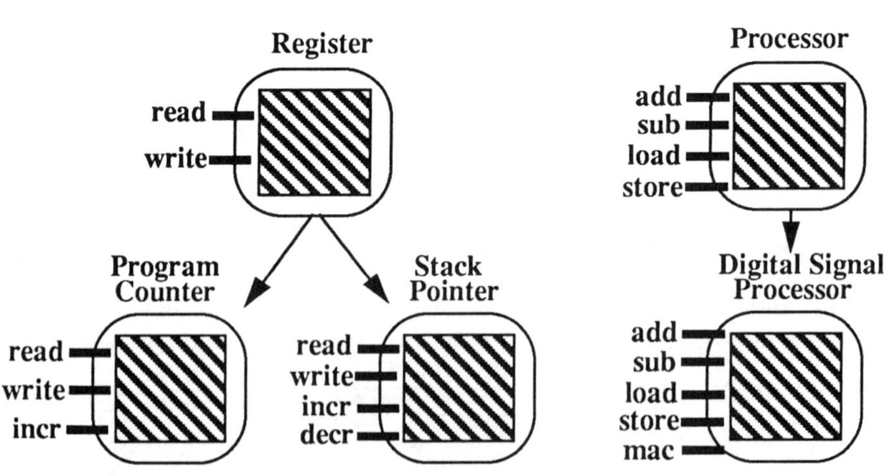

Figure 9.4 Deriving specialized components

Deriving specialized processors suited for a particular application is commonly performed in embedded systems design [276]. To support such specialization, several design automation tools [161][277][278] have been developed for creating application specific processors. Through inheritance, this specialization is accomplished in the same manner as described above.

As an example, a processor containing a collection of basic, "core" instructions can be used to derive a more specialized processor, such as a digital signal processor or a graphics processor. Referring to Figure 9.4, in a digital signal processor, the special instructions may include the multiply-add-accumulate (mac) instruction. Graphics processors may contain special instructions for bit manipulation.

The derived class sp_reg (stack pointer register) is shown in Figure 9.5. Although the READ() and WRITE() member functions do not appear in Figure 9.5, it is still possible to manipulate the contents using these functions. In other words, these operations are implicit within sp_reg. Note that the constructor for class sp_reg invokes the constructor for class reg.

In a similar manner, a memory array class with operations read and write can be used to derive other hardware components. For example, a memory module can be derived from this base class. Also, a register file can be derived with some special operations, such as clear and increment.

These examples illustrate the reuse of code, for example, that of class reg, and the ability to create more specialized components (program counter and stack pointer) from general purpose ones (register). Therefore, new classes of components can be composed quickly and easily, reducing the amount of modeling effort and the time required to validate the models. Also, the close conceptual relationship to libraries should be evident from this discussion.

```
class sp_reg: public reg {                  // derive stack pointer from reg
public:
        sp_reg(int size): reg(size)         // initialize class stack pointer
        {
        }
        void INCREMENT()                    // increment stack pointer
        {
            contents++;                     // increment contents
            write_regbits(contents);        // write the bit representation
        }
        int DECREMENT()                     // decrement stack pointer
        {
            contents--;                     // decrement contents
            write_regbits(contents);        // write the bit representation
        }
}
```

Figure 9.5 Derived register class: stack pointer

9.5 Data Decomposition

Software developers utilize *data decomposition* as a means of refining (deriving implementations for) abstract data types. This section provides an overview of data decomposition and discusses how this technique can be used to identify reusable elements. When modeled as data abstractions, hardware components can also be refined using this decomposition technique. The application of data decomposition to hardware components is explored further in the next section.

9.5.1 Overview of the Technique

In a system developed using functional decomposition, the functions contain intimate knowledge of the data structures. Therefore, extensive modifications can result due to changes in the data structures. Also, because of the tree-like structure of the decomposition, identification of common, reusable functions is difficult.

As illustrated in Figure 9.6, data decomposition is a different refinement technique that manages complexity and addresses the problems mentioned above. In this approach, a high level abstract data type (ADT) is decomposed into a collection of more primitive ADTs whose operations are invoked by one or more transformation functions (see [98] also), that is, algorithms. This decomposition technique can be applied recursively to each of the primitive ADTs.

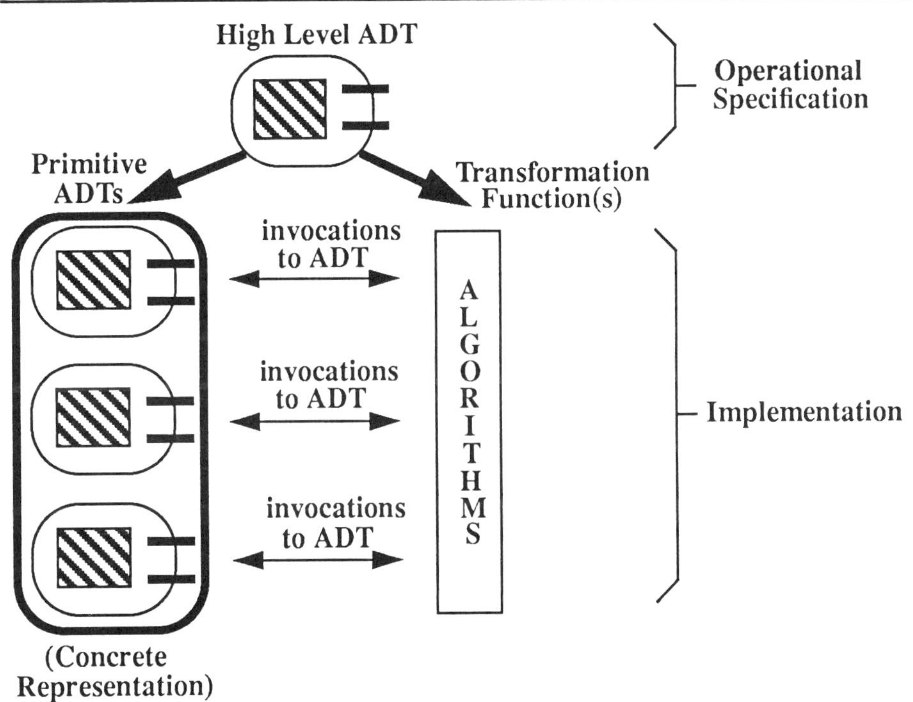

Figure 9.6 Data decomposition

More precisely, data decomposition is equivalent to deriving an implementation for the operational specification of the high level ADT. An implementation for a data abstraction consists of a concrete representation for the data type and implementations for its associated operations. In a data decomposition, the primitive ADTs constitute a concrete representation for the high level ADT. The operations of the high level ADT are implemented in terms of an algorithm that invokes the operations defined on the primitive ADTs.

Referring to equations (9.1)-(9.3), a data decomposition of an ADT can be described as a decomposition graph *DG* consisting of a set of nodes *N* and edges *E*. The decomposition takes the form of a directed acyclic graph (DAG). Each of the nodes $n_k \in N$ corresponds to an ADT. An edge $e_{ij} \in E$ within *DG* represents an ordered pair (n_i, n_j), where n_i "consists of" n_j. In Figure 9.6, the high level ADT consists of the three primitive ADTs shown. As in a *DG* representation of a functional decomposition, the algorithm associated with a data decomposition is implicit within the representation.

$$DG = (N, E) \tag{9.1}$$

$$N = \{n_1, n_2, \ldots, n_z\} \tag{9.2}$$

$$E \subseteq N \times N, \quad (n_i, n_j) \leftrightarrow n_i \textit{ consists of } n_j \tag{9.3}$$

9.5.2 Identifying Reusable Components and Managing Change

Most complex systems are hierarchical, being composed of a common collection of "subsystems" at each level [279]. This idea applies to both software and hardware. In software, common data types may be employed, such as lists, stacks, and matrices. In hardware, the common "subsystems" may be registers, multiplexers, and arithmetic logic units. At a lower level, AND, OR, and XOR gates may be utilized.

Data decomposition can be used to expose these common subsystems. As illustrated in Figure 9.7, the data decomposition of several different ADTs (in this example, two different ADTs *A* and *B)* "points to" a common collection of more primitive ADTs. This ability to identify common, reusable elements stems in part from the DAG structure of the decomposition in which a node can have multiple predecessors. For example, the data decomposition of two processors being designed by different teams may reveal common, reusable components, such as arithmetic logic units, multiplexers, caches, and register files.

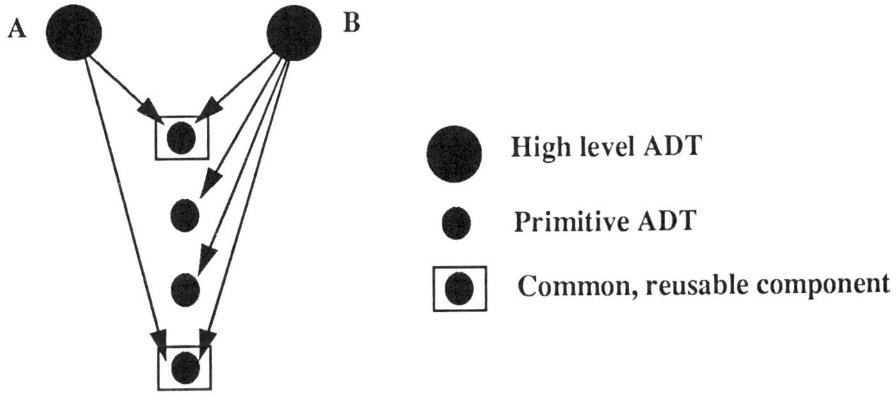

Figure 9.7 Identifying common, reusable components

Change management concerns are also addressed using this approach. Specifically, change is localized to an ADT. As long as implementation changes do not affect the operational specification of an ADT, transformation functions which use the ADT do not require change. Thus, this approach allows reusable components to be identified and supports change.

9.6 PROCESSOR EXAMPLE

In this section, the concepts and ideas presented earlier, particularly data decomposition, are demonstrated on a processor example. Before demonstrating these techniques, the processor's instruction set architecture is described.

9.6.1 Instruction Set Architecture of the Processor

The instruction set architecture of the processor is based on one developed by Williams [280]. The processor is a load-store machine, requiring operands to be placed into registers before any arithmetic operation is performed. The word size is 16 bits, and arithmetic is in two's complement. The set of user-programmable registers is shown in Table 9.1. The condition code register contains three bits that are set by the results of all arithmetic and logical operations. The three bits are set if the result of the operation produces a carry, negative, or zero.

Table 9.1 User-programmable registers

Name	Description
R0-R31	32 16-bit registers
PC	16-bit program counter
CCR	3-bit condition code register
SP	4-bit local stack pointer
STK	16-entry, 16-bit local stack

The instruction set consists of 22 instructions which require either one or two 16-bit words. As indicated in Table 9.2, these instructions are divided into four categories: data movement, control flow, data manipulation, and general. The data movement instructions are used to transfer information between registers, the stack, and memory. The control flow instructions allow for unconditional jumps, conditional jumps, and calls to and returns from subroutines. Both arithmetic and logical operations are provided in the data manipulation category.

Table 9.2 Instruction set for example processor

Data Movement	Control Flow	Data Manipulation	General
MOV, PSH, POP, LDR, STR	JMP, JPC, CSR, RET	ADD, SUB, LAND, LOR, LNOT, LRS, LLS, ARS, RLC, RRC	CLR, CPR, NOP

Up to this point, the instruction set architecture description of the processor has been presented somewhat informally. Conceptually, the instruction set architecture of a machine is a specification which serves as a contract between those who write programs and compilers for the machine and those who implement the machine. This view of a specification is no different than a specification for other abstract data types. Since it is possible to represent processors as abstract data types, formal specifications for processors can be expressed in a similar manner. Of course, this idea can be extended to other hardware components as well.

9.6.2 Data Decomposition of the Processor

Given an operational specification of a processor, an implementation can be derived in terms of more primitive classes of hardware components and a transformation function. Several possibilities exist. For example, the transformation function may consist of a fetch() and an execute() procedure, and the classes would correspond to hardware components, such as a register file, a hardware stack, and an arithmetic logic unit. The transformation function invokes the member functions of these classes. The execute() procedure is described as a collection of case statements, one for each instruction in the instruction set. Additional operand fetching is handled within the branches of the case statement.

The procedure for fetching instructions from a memory is shown in Figure 9.8. This procedure places the program counter (PC) into the memory address register and obtains an instruction from memory. The instruction is then placed into the memory data register and transferred into the instruction register (IR). Finally, to prepare for the next instruction, the program counter is incremented.

```
fetch()
{
        int contents = pc.READ();          // read contents from PC
        mar.WRITE(contents);               // write contents into MAR
        addr = mar.READ();                 // send address to memory
        BitString contents = mem[addr].read_bitrep();    // get instruction
        mdr.write_bitstring(contents);     // write instruction into MDR
        int data = mdr.READ();             // read contents of MDR
        ir.WRITE(data);                    // write contents into IR
        pc.INCREMENT();                    // increment PC
}
```

Figure 9.8 Fetching an instruction

As an example of modeling instruction execution, the statements required to perform a JMP (unconditional jump) instruction are shown in Figure 9.9. Because the JMP instruction has a second word that contains the destination address, this address needs to be fetched. Once the destination address is fetched, the address is placed in the program counter. The code shown represents one "case" of the execute() procedure. If the processor's operational specification consisted of specifications for each instruction, this code would be used to implement the JMP specification.

```
case: JMP
    int contents = pc.READ();          // read contents from PC
    mar.WRITE(contents);               // write contents into MAR
    addr = mar.READ();                 // send address to memory
    bitrep = mem[addr].read_bitrep();  // get jump destination addr.
    mdr.write_bitstring(bitrep);       // write destination into MDR
    addr = mdr.READ();                 // read contents of MDR
    pc.WRITE(addr);                    // write jump address into PC
    break;
```

Figure 9.9 Executing a JMP instruction

9.6.3 Data Decomposition of the ALU

As noted earlier, data decomposition can be continued recursively down to any level of detail, with each abstract data type serving as an operational specification for a lower level implementation. As illustrated in Figure 9.10, an arithmetic logic unit (ALU) can be decomposed into a collection of one-bit arithmetic logic units with some additional gates (not shown) and a corresponding transformation function. The one-bit arithmetic logic units can then be further decomposed into a collection of AND gates, OR gates, inverters, a one-bit full adder, and multiplexers along with a transformation function, and so on. The reusable elements are the classes employed at a given level.

To further illustrate the decomposition technique, consider the ALU of the processor. The operations defined on the ALU consist of add and subtract along with those used to perform logical operations and shifting. Each operation can be implemented as a sequence of invocations to more primitive classes of objects. As shown in Figure 9.11, the add operation (ADD_OP) accepts four inputs represented as bit strings and produces a result which is also a bit string. The first two inputs correspond to the operands to be added. The left and right inputs are bits that are employed during shift operations. This description of the add operation provides a specification for one who wishes to use the add operation of the ALU, isolating the user from implementation changes. As the design of the ALU progresses, common components can be identified as reusable entities.

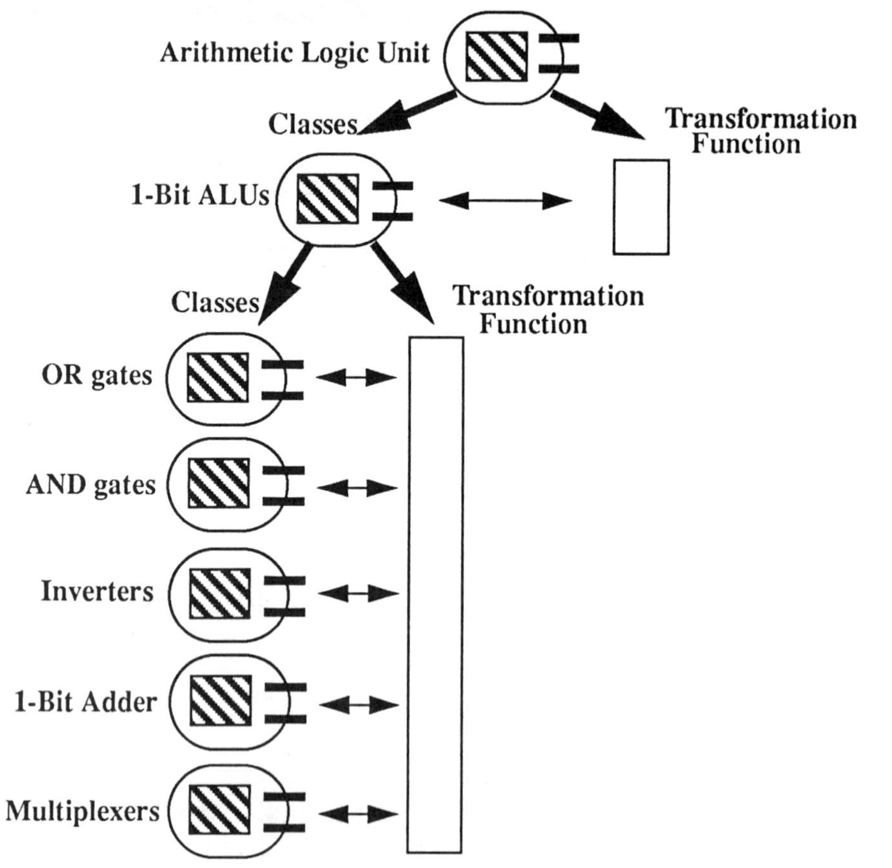

Figure 9.10 Data decomposition of an arithmetic logic unit

```
class arithmetic_unit {                         // class description for ALU
      BitString op1, op2;
      BitString left, right, result;
      int num_bits;
public:
      arithmetic_unit(int size_alu)             // constructor for ALU
      {
                  num_bits = size_alu;
      }
      BitString ADD_OP(BitString op1, BitString op2, \
                        BitString left, Bitstring right)
      {
                  // code for performing add
      }
      BitString SUB_OP(BitString op1, BitString op2, \
                        BitString left, BitString right)
      {
                  // code for performing subtract
      }
      // other operations supported by ALU defined similarly
}
```

Figure 9.11 Portion of class definition of an arithmetic logic unit

Suppose that it is decided to implement the ALU using an interconnection of one-bit ALUs and some miscellaneous gates. This design decision is reflected in the description of the ALU shown in Figure 9.12. The function new() in the constructor allocates memory for the one-bit ALU array at the time that an object is instantiated and returns a pointer to the array.

A portion of the transformation function used to implement the operation ADD_OP is shown in Figure 9.12. This code invokes the operations eval_alures() and eval_alucout(), which compute the sum and carry out, respectively, within a 1-bit ALU. Similar transformation functions are used to implement the other operations of the ALU.

```
class arithmetic_unit {                         // Implementation for ALU
        multiplexer mux1;                       // multiplexer
        or_gate or1;                            // OR gate
        onebit_alu *alu_array;                  // array of 1-bit ALUs
        int num_bits;
        BitString result;
public:
        arithmetic_unit(int size_alu, int size_mux, int size_or): \
        mux1(size_mux), or1(size_or)
        {
            num_bits = size_alu;
            alu_array = new onebit_alu[size_alu]; // array of 1-bit ALUs
        }
        BitString ADD_OP(BitString op1, BitString op2, \
                         BitString left, BitString right)
        {
           // variable declarations & some code here (not shown)
           // main loop
           for (int i= op1.length() - 1; i>= 0; i--) {
                   xinput[i] = op1[i];
                   yinput[i] = op2[i];
                   if (i == 0)
                       xlinput = left;
                   else
                       xlinput = op1[i-1];
                   tmp = alu_array[i].eval_alures(xlinput, xinput, xrinput, \
                                              yinput, carry_in, alu_oper);
                   tmp_result += tmp;
                   carry_in = alu_array[i].eval_alucout(xlinput, xinput, \
                                              xrinput, yinput, carry_in, \
                                              alu_oper);
                   xrinput = xinput;
           }
           // some code here (not shown)
            return(result);
        }
        // other operations supported by ALU defined similarly
}
```

Figure 9.12 Portion of the ALU implementation

At this point, the 1-bit ALU can be decomposed into more primitive elements (see Figure 9.10). The operational specification for one of the elements, a one-bit full adder, is shown in Figure 9.13. The one-bit adder can be concretely represented as a collection of XOR gates, AND gates, and an OR gate. The operations performed on the one-bit adder include evalsum() and evalcout(), which would be implemented as sequences of invocations to these more primitive classes. One possible implementation of the one-bit adder is shown in Figure 9.14.

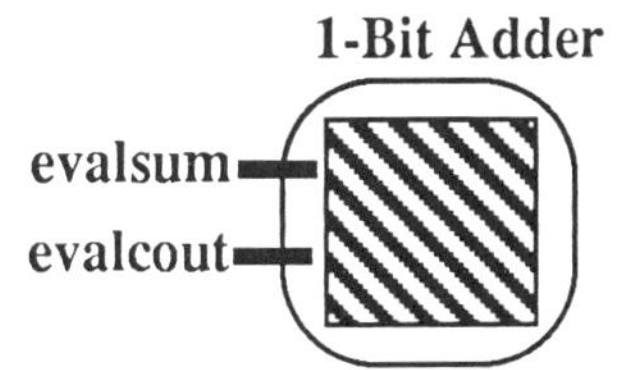

$evalsum(BitString\ x, BitString\ y, BitString\ cin)$
$returns\ BitString$
$pre\ length(x) = length(y) = length(cin) = 1$
$post\ result \leftarrow x \oplus y \oplus cin$

$evalcout(BitString\ x, BitString\ y, BitString\ cin)$
$returns\ BitString$
$pre\ length(x) = length(y) = length(cin) = 1$
$post\ result \leftarrow xy \vee xcin \vee ycin$

Figure 9.13 Operational specification for a one-bit full adder

Note that existing software verification techniques [111][129][281] can be applied to hardware components. Because the processor is modeled using data abstractions, techniques for verifying the correctness of abstract data type implementations with respect to their operational specifications can be utilized. By recursively applying these techniques, the correctness of the processor implementation can be established with respect to its operational specification.

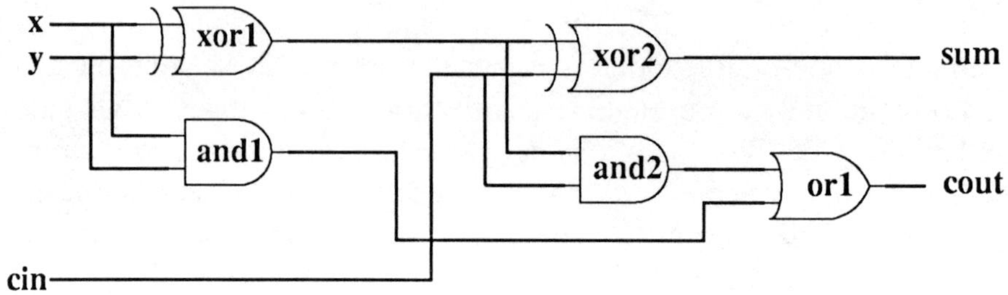

```
class onebit_adder {                    // class description for onebit_adder
        and_gate and1;
        and_gate and2;
        xor_gate xor1;
        xor_gate xor2;
        or_gate or1;
        BitString sum;
        BitString carry_out;
public:
        onebit_adder()                  // constructor for onebit_adder
        {
            sum = atoBitString("0");
            carry_out = atoBitString("0");
        }
        BitString evalsum(BitString x, BitString y, BitString cin)
        {
            BitString out_xor1 = xor1.evaluate(x + y);
            BitString sum = xor2.evaluate(out_xor1 + cin);
            return(sum);
        }
        BitString evalcout(BitString x, BitString y, BitString cin)
        {
            BitString out_xor1 = xor1.evaluate(x + y);
            BitString out_and1 = and1.evaluate(x + y);
            BitString out_and2 = and2.evaluate(out_xor1 + cin);
            BitString carry_out = or1.evaluate(out_and2 + out_and1);
            return(carry_out);
        }
}
```

Figure 9.14 One-bit adder implementation

9.6.4 Inheritance Hierarchies

A description of the inheritance hierarchies employed in this model is depicted in Figure 9.15. Each of the nodes in the hierarchy designates a class. The top node in each hierarchy is the base class, and the lower nodes are derived classes. For example, the class gate is used to derive the classes AND gate, OR gate, XOR gate, and inverter. In the same way, a register class is used to derive a program counter class, a stack pointer class, and an instruction register class.

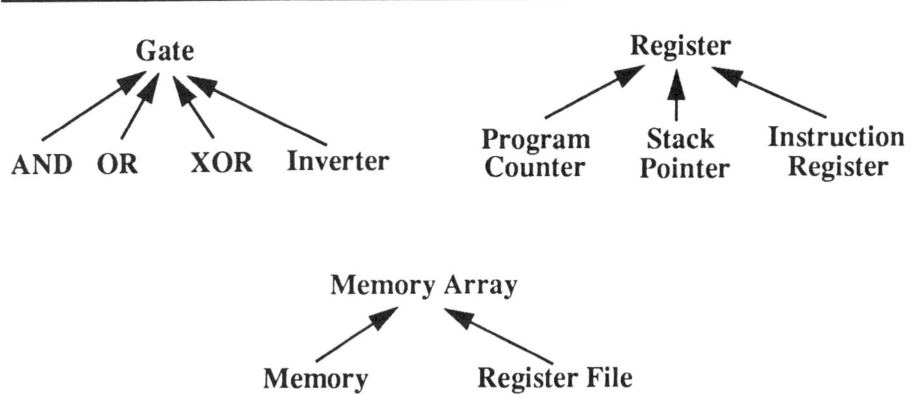

Figure 9.15 Inheritance hierarchies employed

Although not developed as part of the model, there are other classes of components that are useful. Flip-flops and latches are fundamental sequential devices used in the design of digital systems. Thus, one can construct a class called flip-flop and derive D flip-flops, J/K flip-flops, and S/R flip-flops. Although the model did not develop the control unit for the processor, the register class could have been used to derive the microprogram counter, and a control memory for storing microinstructions could have been derived from the memory array class (assuming a writable control store). Also, specialized devices can be derived from existing classes depending upon the application. As an example, application specific processors may utilize specialized arithmetic logic units.

9.6.5 Discussion

Some points are worth mentioning about a model developed using data decomposition. The resulting model resembles a data path/control decomposition. When a processor is instantiated, the components constitute the elements of the data path, and the transformation function represents the control, that is, the sequencing of the operations to be performed. Each machine instruction is implemented as a collection of invocations to the components required for execution. As an illustration, in order to fetch and execute a subtract instruction, operations are performed on the memory address register, the memory data register, the register file, the arithmetic logic unit, the condition code register, and the program counter. Thus, the transformation function represents the register transfers to be initiated by a microcoded or a hardwired control unit.

Data decomposition illustrates a fundamental idea within the context of virtual machines. Specifically, a virtual machine at level *N* is implemented in terms of a program running on a machine at level *N-1* [282]. For instance, the ISA level model of the processor can be considered a virtual machine which is implemented as a microprogram running on a lower level machine (micromachine), consisting of arithmetic logic units, register files, and so on. One can extend this analogy to lower levels of the virtual machine hierarchy. From a conceptual standpoint, each abstract data type (class) may be considered a virtual machine, a reusable element with a corresponding virtual instruction set. Thus, using this technique, a virtual machine is decomposed into more primitive virtual machines, each of which "executes" a portion of an abstract program.

Note that the abstract data types within a data decomposition do not necessarily have to coincide one-to-one with our intuitive notion of a component. The abstract data types utilized at any given level may be based on complexity considerations. Consider a one-bit full adder. A full adder can be expressed directly in terms of AND gates, XOR gates, and an OR gate. Alternatively, a full adder may be implemented in terms of half adders and an OR gate. Then, at the next level, the half adder class can be expressed in terms of an AND gate and an XOR gate. Although this is a simple example, the point being made is that the abstract data types employed to implement a component depend on the amount of complexity that can be managed at that level.

9.7 TYPE GENERICITY

Type genericity [283], a form of *polymorphism* [284][285], is the ability to parameterize a software element, such as a procedure or a data type, with one or more types. Type genericity makes programs more general. To illustrate this point, a swap procedure can be parametrized with a type, allowing either two integers, two reals, or two strings to be swapped. In the same manner, data types can be parameterized with types as well. Thus, one can have a data type called list which is parameterized with another type. Using the three types mentioned above, it is possible to create a list of integers, a list of reals, or a list of strings. Several languages, such as Alphard [129] and CLU [117], support this parameterization of types. This kind of polymorphism is particularly useful when dealing with objects which act as "containers" for other elements, such as queues, lists, and arrays.

In hardware, several components, such as registers, stacks, and register files, act as containers. Thus, as shown in Figure 9.16, one can utilize templates in C++ to construct a hardware stack which stores integers or floating point numbers. In the same manner, a register file class can be created to represent an integer register bank or a floating point register bank. Also, memory stores instructions, which can also be treated as a type. One of the benefits of type genericity is reuse of code. The same code can be employed to construct containers for different data types.

```
main()
{
        stack<int> stk1(10);        // stack of 10 integers
        stack<float> stk2(15);      // stack of 15 floats

        stk1.push(3);               // push the integer 3 onto the integer stack
        stk2.push(4.5);             // push the float 4.5 onto the float stack
}
```

Figure 9.16 Type genericity

The notion of type genericity is closely associated with Parnas's concept of program families [286]. A program family (or design family) corresponds to a set of programs with common characteristics. The hierarchical structuring of program families addresses design change and can help reduce the cost of design development and maintenance. The fundamental idea behind program families is to emphasize those properties that are common to a family of designs before considering special properties which distinguish between individual family members. This idea can also be applied to the development of families of hardware components, such as families of processors.

9.8 Related Work

Previous work in the application of object-oriented concepts to hardware modeling and design has appeared in several forms. Giloi [287] presents a taxonomy of computer architectures using machine data types, which are based on descriptions of abstract data types. Langdon [288] has noted the active-passive duality which exists throughout digital systems and has related this idea to elements within a computer system. He describes the decomposition of problems into an active part and a passive part. Within a processor, components such as registers and arithmetic units constitute the objects (passive part), and control (active part) provides the actions which change the state of the objects.

Gross [181] addressed the problem of change management concerns in the VLSI design process using Parnas's information hiding principle. Müller and Rammig [289] describe a hardware description language which models hardware components as classes and supports single inheritance. Chung and Kim [290] have also used class concepts for describing VHDL design entities. Nelson, Fontes, and Zaky [291] have developed "generic" classes of hardware components to aid in the modeling and simulation of computer architectures at the microarchitecture level, as well as families of computer architectures. More recently, object-oriented extensions to VHDL have been developed [292] to support rapid prototyping, performance modeling, and hardware/software codesign. These extensions incorporate the capability of expressing components as classes and employing inheritance.

The use of object-oriented concepts and ideas has also been discussed in system modeling and electronic design automation systems. Newton, Vaughn, and Johns [293] and SES/Workbench [76] have used these ideas within the context of queuing-based system level modeling environments. Yokote and Tokoro [294] have utilized object-oriented design techniques for modeling distributed and fault-tolerant systems. More generally, several individuals [285][295][296] have promoted the use of object-oriented techniques in electronic design automation.

Ledbetter and Cox [297] discuss the notion of software-ICs (integrated circuits). They describe the need to incorporate ideas associated with hardware reusability into software. Some of these ideas include developing appropriately "packaged" components that provide well defined services and hide the internal operation of an IC. The concept of IC foundries, which produce standard and custom ICs, is also promoted.

The research described in this chapter is different from previous object-oriented hardware modeling efforts in several respects. The focus of this effort has been to develop a unified view of hardware and software through the use of common modeling and refinement techniques. It has been shown that data abstractions, commonly used in software design, can be used to model hardware. Through data decomposition, it has been demonstrated that hardware models can be represented and refined using the same set of techniques that are commonly used in the software domain. In particular, the specification and implementation of abstract data types can be applied in a uniform fashion to both hardware and software. With regard to reuse, one can employ models that are common to both hardware and software, such as queues and stacks.

Several insights have been gained as a result of modeling hardware components through data abstraction. By modeling hardware components as abstract data types and employing data decomposition as a refinement technique, the conceptual model of virtual machines is readily exposed. In such a model, abstract data types can be considered virtual machines whose operations correspond to a virtual instruction set. Therefore, in the same manner that a functional decomposition defines a virtual machine with a virtual instruction set, a data decomposition does so as well. Also, type genericity in hardware modeling and design supports Parnas's concept of program families.

9.9 SUMMARY

This chapter has presented and illustrated several potential advantages of employing object-oriented techniques in hardware modeling and design. Some of these advantages include complexity management, specialization, and model reuse. Object-oriented concepts provide new techniques for hardware development. Specifically, in the process of refining a hardware component, data decomposition, the decomposition of abstract data types into more primitive abstract data types, supports the identification of common, reusable hardware components and change management.

More importantly, within the context of hardware/software codesign, object-oriented techniques provide a unified way of viewing both hardware and software. A unified model of hardware and software allows techniques from one domain to be applied to the other. The discussions presented in this chapter have further blurred the distinction between hardware and software. Beyond the physical characteristics associated with hardware, it is apparent that the "difference" between hardware and software is only one of level within the virtual machine hierarchy.

Some aspects of hardware design have not been discussed. For example, given a specification of a hardware component described as an abstract data type, it may be possible to use software synthesis techniques [140][298] as a means of generating an implementation. Also, the aspect of concurrency, which is fundamental in hardware modeling, is addressed elsewhere [292][294].

Chapter 10
Concluding Remarks and Future Work

In this chapter, the important ideas of the monograph are summarized. Directions for future work are outlined. Finally, concluding remarks are provided.

10.1 MONOGRAPH SUMMARY

This section contains a summary of the important ideas from Chapter 4 through Chapter 9. The discussion starts with Chapter 4 and ends with Chapter 9.

10.1.1 Codesign Concepts

Several definitions were introduced to serve as a common base for discussions in this monograph. One idea that emerged from this chapter was the quantitative evaluation model, a linear, weighted model which could be used to assess the quality (goodness) of a hardware/software alternative with respect to multiple metrics, such as performance, cost, and reliability. Using this evaluation model, it was demonstrated that varying the weights associated with certain metrics influences the quality of a hardware/software alternative.

10.1.2 A Methodology for Codesign

The monograph promotes the idea that a codesign process is different from the traditional hardware/software design process. Two distinguishing characteristics of a codesign process include the utilization of a unified representation and support for iterative hardware/ software partitioning. A codesign methodology was discussed that incorporates both of these characteristics. In addition, the methodology supports incremental evaluation and the movement of functionality from one domain to the other.

This approach contrasts with many existing approaches which permit only instruction set level evaluation. These existing approaches force the designer to describe the system at a particular level of representation (typically the instruction set level) before allowing any evaluation to occur. Also, some of these approaches project an overly simplistic view of the hardware/software design process, neglecting the complexity of both the design and analysis of systems.

10.1.3 A Unified Representation for Hardware and Software

A unified representation, referred to as a decomposition graph, was introduced in which the nodes correspond to either functional abstractions or data abstractions. This representation can be used to describe software and hardware. The decomposition graph incorporates the modeling concepts of abstraction level and interpretation level, two aspects that are shared by hardware and software models.

There are several benefits to be derived from the use of such a representation. Specifically, opportunities exist for cross fertilization from one domain to the other. For example, common techniques can be employed for the determination of performance, reliability, and formal verification of correctness. The representation also can serve as a starting point for hardware and software synthesis. Finally, a unified representation provides a common modeling paradigm, one that can be understood by both hardware and software developers.

10.1.4 An Abstract Hardware/Software Model

An abstract hardware/software model employing a unified representation was described. The unified representation is based on functional abstractions and utilizes data/control flow concepts. This modeling approach permits hardware and software to be developed cooperatively in a common simulation environment. In addition, the model supports evaluation at different levels of detail, providing a designer the flexibility of focusing on those aspects of interest while ignoring others. The use of appropriate abstractions allows alternatives to be evaluated more quickly, with respect to multiple metrics.

The abstract hardware/software model has many applications. The model supports early evaluation, allowing the consequences of hardware/software decisions to be assessed before committing to a particular implementation. In addition to general performance evaluation, the model can be used to identify software bottlenecks, evaluate hardware/software trade-offs, and evaluate design alternatives. The model was implemented in the ADEPT environment, and the applications of the model were demonstrated on several examples. Chapter 7 focused on general performance evaluation. Chapter 8 emphasized bottleneck analysis, trade-off evaluation, and alternative evaluation. The post-processing programs *opsens* and *alteval* were used to support software bottleneck analysis and hardware/software alternative evaluation, respectively. The latter implements the quantitative evaluation model.

The integration of both hardware and software descriptions within a common environment also supports combined performance and reliability evaluation at (perhaps) several stages of the design process. Thus, another possible application of the model is that of an integrated substrate. By utilizing the hardware/software model as an integrated substrate, evaluation and trade-off exploration can be performed incrementally, as hardware and software are developed. Lower level implementations can also be incorporated into the model, supporting model continuity.

In addition to the notion of a common representation, another unifying idea embodied within the abstract hardware/software model is that of an interpreter. The concept of an interpreter is common to both hardware and software. By viewing the hardware model as a virtual machine abstraction (software or hardware interpreter), a more general model results, one which represents interactions between software developers or between software and hardware developers.

10.1.5 Object-Oriented Techniques in Hardware Design

A decomposition graph consisting of nodes which correspond to data abstractions was explored as a unified representation for hardware and software. Using a processor example, this work illustrated how concepts and techniques from the software domain could be applied to the hardware domain (cross fertilization). It was shown that data abstractions could be used to model hardware. Through data decomposition, it was demonstrated that hardware models could be refined using the same techniques utilized in the software domain. A common representation allows existing software verification techniques to be used for hardware, further supporting the idea of cross fertilization. Thus, the specification, implementation, and verification of data abstractions can be applied in a uniform fashion to both hardware and software.

Inheritance and genericity were also investigated within the context of hardware. Inheritance was employed to create more specialized hardware components. An inheritance hierarchy was constructed for the processor model to highlight the ideas of reuse and the common properties shared by typical hardware components. The concept of genericity, particularly type parameterization, further illustrated code reuse and the notion of program families.

Conceptually, data abstractions can be considered virtual machines whose operations correspond to a virtual instruction set. As a result, in the same manner that a functional decomposition defines a virtual machine with a virtual instruction set, a data decomposition does so as well. Therefore, beyond the physical characteristics associated with hardware, one can conclude that the "difference" between hardware and software is only one of level within the virtual machine hierarchy.

10.2 FUTURE WORK

The implementation of the abstract hardware/software model was an important first step in supporting a codesign capability within the ADEPT environment. However, there are several possible directions for future work.

10.2.1 Hardware/Software Partitioning Algorithms

An area that has been given only cursory treatment in this monograph is the development of algorithms for performing hardware/software partitioning. The search for automatic hardware/software partitioning algorithms continues to be one of the more prominent research problems in codesign. Although the ADEPT environment supports manual hardware/software partitioning, better approaches are necessary.

One possibility is to investigate approaches based on the utilization of software and hardware elements that can be "swapped" in place [299]. Another possibility is to employ knowledge-based approaches that incorporate domain-specific information. A more ambitious effort involves the development of adaptive approaches. These approaches may be coupled with the use of cost tables [199], which tabulate the cost of implementing functionality in software versus hardware. Note that the quantitative evaluation model can be integrated with a hardware/software partitioning algorithm in a tight "loop", providing quick assessment of the partitioning decision.

10.2.2 Hardware/Software Trade-offs

It is desirable to determine under what circumstances a hardware/software trade-off will provide a benefit. As an example, consider the movement of functionality from software to hardware. If the new hardware operation is implemented using only existing resources, the execution of other operations may not be affected. However, if the new hardware operation requires existing resources plus some additional ones (or additional functions), this movement of functionality into hardware may adversely affect the execution of other operations due to increased delays in the data path. As a result, the overall performance may be degraded. In some circumstances, a new execution unit may be developed in which case additional decode logic will be required.

Some rigorous techniques are required to guide the process of performing trade-offs. One possible heuristic is to use the notion of operator similarity [151]. In the example above, a new operation which has a high similarity with existing operations implemented in hardware may be less likely to adversely affect the execution of these operations.

Another area that deserves further investigation is the development of quantitative expressions which capture the cost of implementing functionality in hardware versus software. A good starting point for exploring this area is the discussion by Myers [146]. Myers quantifies the cost of implementing functionality in software and in hardware by considering both development cost and manufacturing cost. Economic issues, such as the number of units produced and the number of software "copies" developed, are factored into the cost expressions. Along the same lines, it would be interesting to explore a "unified" cost model, one which would take advantage of the similarities between hardware and software.

10.2.3 Hardware/Software Synthesis

The unified representation in the abstract hardware/software model has been used for performance analysis. Additional work is necessary to investigate the use of the unified representation for hardware and software synthesis. In general, representations based on either functional abstractions or data abstractions can serve as a starting point for synthesis. Formal techniques can be incorporated into these capabilities to verify the correctness of the final implementation.

10.2.4 Modeling and Analysis

There are several possible extensions to the abstract hardware/software model. It would be worthwhile to utilize the model for combined hardware/software reliability analysis. For example, Markov models can be extracted to estimate the overall reliability. One attempt at exploring such an approach is found in [300].

The abstract hardware/software model developed in this monograph focuses on the evaluation of a single program on a single processor. Models need to be developed which can support the analysis of more complicated parallel and distributed systems. Most likely, this effort will require a different set of primitive modules. These models must be robust enough to analyze the partitioning and scheduling (static and dynamic) of programs across multiple processors while utilizing different interprocessor communication strategies. It would also be useful to explore "mixed level" simulations of parallel and distributed systems in which some portions are modeled using the abstract hardware/software model while others are even more abstract.

The abstract hardware/software model can be employed within a hybrid model in two different ways. First, a lower level (interpreted) hardware component can be incorporated into the hardware model. Alternatively, the entire abstract hardware/software model can be treated as the interpreted element. The former approach represents the current view of hybrid modeling. However, the latter approach is possible as well.

Although the use of the abstract hardware/software model as an integrated substrate has been discussed, more work is required to bring this idea to fruition. One task is the compilation of a source language into an intermediate form. The process of generating intermediate representations for virtual machines can be applied here [251]. Of course, additional work is necessary to support incremental evaluation and trade-off exploration as the hardware/software development proceeds. This methodology assumes a flexible design approach in which hardware and software models can be generated quickly.

The work on applications of object-oriented techniques can be extended in several ways. Object-oriented techniques may be applied to hardware design as well as system level modeling. Also, synthesis and verification tools can be developed based on data abstractions (see [298] and [129]). Thus, it is feasible to develop a unified design environment based on these ideas. To initiate investigations in this area, it may be worthwhile to look at object-oriented system level modeling capabilities and object-oriented extensions to VHDL. As a final note, the object-oriented work may be explored further using Mentat [301], an object-oriented parallel programming environment, particularly the modeling of parallel and distributed systems.

10.2.5 Formal Techniques

There are many applications of formal techniques. In some circumstances, it is desirable to support verification of correctness. The incorporation of formal specifications is necessary to support such a capability. Another benefit is that timing can be incorporated into these descriptions as well [281]. It may also be possible to develop algebras for composing hardware/software systems from a set of building blocks. Some work in this area can be found in [302]. These techniques can be integrated into synthesis tools.

10.3 CONCLUDING REMARKS

It is believed that a cooperative design approach, one that utilizes a unified view of hardware and software, addresses the deficiencies of the existing development process and thus, benefits the design of complex systems. This idea was supported through the concept of a unified representation and the notion of an abstract hardware/software model. The unified representation provides a common view of hardware and software. The abstract hardware/software model permits the unification of the two domains and thus, allows cooperative hardware/software design.

An attempt has been made to unify the currently separate hardware and software domains by recognizing the similarities between the two. Many of these similarities are reflected in the decomposition graph. This unified representation allows hardware or software to be described through either functional abstractions or data abstractions. Complexity management through information hiding, an idea important to both hardware and software, is a fundamental element of the representation. The notion of virtual machines is inherent as well. Also, the modeling concepts of abstraction level and interpretation level are common aspects that are embodied within the representation. Although interpretation level has been described within the context of functional abstractions, this concept can be applied to data abstractions as well.

The abstract hardware/software model also promotes the idea of unification between the two domains. The model utilizes a unified representation based on functional abstractions, employing data/control flow concepts. The notion of an interpreter, another idea shared by both hardware and software, has been incorporated into the model. A more general form of the model allows one to represent interactions between software developers or between software and hardware developers. In addition, the model's (potential) use as an integrated substrate supports continuous hardware/software integration and evaluation.

The abstract hardware/software model supports cooperative design as well. A cooperative design approach can reduce costs by preventing late changes to hardware and/or software. Using this model, early performance evaluation and trade-off exploration are possible. The use of hardware/software abstractions addresses the time consuming execution of instruction set level descriptions. These abstractions allow systems to be evaluated at different levels of detail with respect to multiple metrics.

The utility of these ideas was illustrated through several examples. The benefits of a unified representation were demonstrated in primarily two ways. A representation employing functional abstractions was used to support integrated hardware/software performance analysis. A representation based on data abstractions showed how object-oriented techniques could be applied to hardware design, supporting the notion of cross fertilization.

Through the implementation of the abstract hardware/software in ADEPT, several benefits of the model were demonstrated. Using a unified representation, it was shown that several types of hardware/software evaluation were possible in an integrated design environment: general performance evaluation, the identification of bottlenecks, the evaluation of hardware/software trade-offs, and the evaluation of design alternatives. The aspect of cooperative design was most clearly demonstrated through the process of performing trade-offs, particularly the movement of functionality from one domain to the other.

As a final note regarding the abstract hardware/software model, the issue of accuracy was not addressed since this determination was left to the model developer. It should be remembered that any model is only as good as the assumptions made during its construction. A model constructed from unrealistic assumptions will not accurately reflect the final system. It is also true that adding more detail will increase the model's accuracy. However, it should be kept in mind that absolute accuracy is not always what is required, particularly during the early stages of the design process. In many circumstances, it is desirable to focus on only those aspects that are of interest or concern.

The investigations in this monograph have blurred the distinction between hardware and software. As a result, systems (or portions of a systems) can be viewed independent of hardware or software. An obvious example is focusing on the function to be performed without regard to hardware or software implementation. Similar reasoning can be applied to certain data types, such as queues and stacks. A more subtle example appears within the context of interpreters. Consider the more general form of the abstract hardware/software model. If the hardware model is viewed as an abstract interpreter, the interpreter can be either a software or a hardware element. Note that this example illustrates the concept of information hiding and represents another type of hardware/software trade-off.

An important theme in this monograph has been the cross fertilization between the software and hardware domains. Cross fertilization is both important and essential to future work in the area of codesign. The cross fertilization of knowledge will help to further unify these currently separate domains.

References

[1] Turn, R., "Hardware-Software Tradeoffs in Reliable Software Development," *11th Annual Asilomar Conference on Circuits, Systems, and Computers*, 1978, pp. 282-288.

[2] Harding, B., "Mixed CASE and CAE/CAD Tools Ease Designers' Headaches," *Computer Design*, January 1, 1990, pp. 74-88.

[3] Franke, D. W., M. K. Purvis, "Hardware/Software Codesign: A Perspective," *Proceedings of the 13th International Conference on Software Engineering*, May 13-16, 1991, pp. 344-352.

[4] Iyer, V. R., H. A. Sholl, "Software Partitioning for Distributed, Sequential, Pipelined Applications," *IEEE Transactions on Software Engineering*, Vol. 15, No. 10, October 1989, pp. 1270-1279.

[5] Peng, Z., J. Fagerström, K. Kuchcinski, "A Unified Approach to Evaluation and Design of Hardware/Software Systems," 1991 Workshop on Hardware/Software Codesign, Technical Report No. MCC-CAD-156-91.

[6] Roman, G., M. J. Stucki, W. E. Ball, W. D. Gillett, "A Total System

Design Framework," *IEEE Computer*, May 1984, pp. 15-26.

[7] Boehm, B. W., "Software and its Impact: A Quantitative Assessment," *Datamation*, May 1973, pp. 48-59.

[8] Smith, C. U., L. G. Williams, "Software Performance Engineering: A Case Study including Performance Comparison with Design Alternatives," *IEEE Transactions on Software Engineering*, Vol. 19, No. 7, July 1993, pp. 720-741.

[9] Boehm, B. W., "Software Engineering," *IEEE Transactions on Software Engineering*, Vol. C-25, No. 12, December 1976, pp. 1226-1241.

[10] Dunn, R. H., *Software Defect Removal*, McGraw-Hill Inc., 1984.

[11] Van Genuchten, M., "Why is Software Late? An Empirical Study of Reasons for Delay in Software Development," *IEEE Transactions on Software Engineering*, Vol. 17, No. 6, June 1991, pp. 582-590.

[12] McFarland, M. C., A. C. Parker, R. Camposano, "The High-level Synthesis of Digital Systems," *Proceedings of the IEEE*, Vol. 78, No. 2, 1990, pp. 301-318.

[13] Patterson, D., C. Sequin, "A VLSI RISC," *IEEE Computer*, September 1982.

[14] Franke, D. W., M. K. Purvis, "Design Automation Technology for Codesign: Status and Directions," *International Symposium on Circuits & Systems*, May 1992, pp. 2669-2671.

[15] Frank, G. A., D. L. Franke, W. F. Ingogly, "An Architecture Design and Assessment System," *VLSI Design*, August 1985, pp. 30-50.

[16] Terry, C., "Concurrent Hardware and Software Design Benefits Embedded Systems," *EDN*, July 1990, pp. 148-154.

[17] Zurcher, F. W., B. Randell, "Iterative Multi-level Modelling - A Methodology for Computer System Design," *Proceedings IFIP Congress '68*, Edinburgh, Scotland, August 1968, pp. 867-871.

[18] Ferrari, D., "Considerations on the Insularity of Performance Evaluation," *IEEE Transactions on Software Engineering*, Vol. SE-12, No. 6, June 1986, pp. 678-683.

[19] Bourbon, B., "On System Level Design," *Computer Design*, December 1990.

[20] Schultz, S. E., "An Overview of System Design," *ASIC & EDA*, January 1993, pp. 12-21.

[21] Aylor, J. H., R. Waxman, B. W. Johnson, R. D. Williams, "The Integration of Performance and Functional Modeling in VHDL" in *Performance and Fault Modeling with VHDL*, J. Schoen, ed., Prentice-Hall, Englewood Cliffs, N. J., 1992.

[22] Hill, D., D. Coelho, *Multi-level Simulation for VLSI Design*, Kluwer Academic Publishers, 1987.

[23] Srivastava, M. B., R. W. Broderson, "Using VHDL for High-level, Mixed-Mode System Simulation," *IEEE Design & Test*, September 1992, pp. 31-40.

[24] Browne, J. C., "Codesign and Codevelopment of Hardware/ Software Systems: Representational Issues," International Workshop on Hardware/Software Codesign, Estes Park, Colorado, September 1992.

[25] Kumar, S., J. H. Aylor, B. W. Johnson, W. A. Wulf, "Hardware/ Software Modeling & Evaluation in a Unified Codesign Environment," Department of Electrical Engineering, University of Virginia, Technical Report No. 920505.0, May 5, 1992.

[26] De Micheli, G., "Extending CAD Tools and Techniques," in Hot Topics of *IEEE Computer*, R. D. Williams, ed., January 1993, pp. 84-87.

[27] Kundig, A. T., "A Note on the Meaning of Embedded Systems," in *Lecture Notes in Computer Science*, G. Goos, J. Hartmanis, eds., Embedded Systems: New Approaches to their Formal Description and Design, A. Kundig, R. E. Buhrer, J. Dahler, eds., Springer-Verlag, 1987, pp. 1-6.

[28] Harel, D., "On Visual Formalisms," *Communications of the ACM*, Vol. 31, No. 5, May 1988, pp. 514-530.

[29] Harel, D., "Biting the Silver Bullet: Toward A Brighter Future for System Development," *IEEE Computer*, January 1992, pp. 8-19.

[30] Benveniste, A., G. Berry, "The Synchronous Approach to Reactive and Real-Time Systems," *Proceedings of the IEEE*, Vol. 79, No. 9, September 1991, pp. 1270-1282.

[31] Cook, R., "Embedded Systems in Control," *Byte*, June 1991, pp. 153-160.

[32] Wilson, R., "Embedded Systems Manipulate Distributed Tasks," *Computer Design*, September 1, 1987, pp. 49-61.

[33] Wilson, R., "Higher Speeds push Embedded Systems to Multiprocessing," *Computer Design*, July 1, 1989, pp. 72-83.

[34] Johnson, B. W., *Design and Analysis of Fault Tolerant Digital Systems*, Addison-Wesley Publishing Company, Inc., 1989.

[35] Bell, C. G., A. Newell, "The PMS and ISP Descriptive Systems for Computer Structures," *AFIPS Conference Proceedings*, Vol. 36, 1970, pp. 351-374.

[36] Bell, C. G., A. Newell, *Computer Structures: Readings and Examples*, McGraw-Hill Book Company, New York, 1971.

[37] Gajski, D., R. Kuhn, "Guest Editors' Introduction: New VLSI Tools," *IEEE Computer*, December 1983, pp. 11-14.

[38] Walker, R. A., D. E. Thomas, "A Model of Design Representation and Synthesis," *Proceedings of the 22nd Design Automation Conference*, 1985.

[39] Peterson, J., "Petri Nets," *Computing Surveys*, Vol. 9, No. 3, September 1977, pp. 223-252.

[40] Ferrari, D., *Computer Systems Performance Evaluation*, Prentice-Hall, Englewood Cliffs, NJ, 1978.

[41] Dugan, J. B., S. Bavuso, M. Boyd, "Dynamic Fault Tree Models for Fault Tolerant Computer Systems," *IEEE Transactions on Reliability*, September, 1992.

[42] Barbacci, M. R., "Instruction Set Processor Specification (ISPS): The Notation and its Applications," *IEEE Transactions on Computer*, Vol. C-30, January 1981, pp. 26-40.

[43] Dewey, Lt. A., "VHSIC Hardware Description (VHDL) Development Program," *Proceedings 20th Design Automation Conference*, June 1983, pp. 625-628.

[44] Aylor, J. H., R. Waxman, C. Scarratt, "VHDL - Feature Description and Analysis," *IEEE Design & Test*, April 1986, pp. 17-27.

[45] Barton, D. L., "Behavioral Descriptions in VHDL," *VLSI Systems Design*, June 1988, pp. 28-33.

[46] Kernighan, B. W., *The C Programming Language*, 2nd Edition, Prentice-Hall, Inc., Englewood Cliffs, NJ, 1988.

[47] Tanenbaum, A. S., *Operating Systems: Design and Implementation*, Prentice-Hall, Inc., Englewood Cliffs, NJ, 1987.

[48] Tanenbaum, A. S., *Structured Computer Organization*, 3rd Edition, Prentice-Hall of India, New Delhi, India, 1991.

[49] Dijkstra, E. W., "The Structure of the T.H.E. Multiprogramming System," *Communications of the ACM*, Vol. 11, No. 5, May 1968, pp. 341-346.

[50] Dijkstra, E. W., "Structured Programming," *Software Engineering Techniques*, NATO Science Committee, 1969, p. 84-88.

[51] Randell, B., "System Structure for Software Fault Tolerance," *Programming Methodology - A Collection of Articles by Members of IFIP WG2.3*, D. Gries, ed., Springer-Verlag, New York, 1968, pp. 362-387.

[52] Mills, H. D., "Software Engineering - Retrospect and Prospect,"

Computer and Software Applications Conference '88, 1988, p. 89-96.

[53] Parnas, D. L., "Designing Software for Ease of Extension and Contraction," *IEEE Transactions on Software Engineering*, Vol. SE-5, No. 2, March 1979.

[54] Zuberek, W. M., "Timed Petri Nets and Preliminary Performance Evaluation," *Proceedings 7th Annual Symposium Computer Architecture*, 1980, pp. 89-96.

[55] Molloy, M. K., "Performance Analysis using Stochastic Petri Nets," *IEEE Transactions on Computer*, Vol. C-31, No. 9, September 1982, pp. 913-917.

[56] Holliday, M. A., M. K. Vernon, "A Generalized Timed Petri Net for Performance Analysis," *IEEE Transactions on Software Engineering*, Vol. SE-13, No. 12, December 1987.

[57] Chang, C. K., Y. Chang, L. Yang, C. Chou, J. Chen, "Modeling a Real-Time Multitasking System in a Timed PQ Net," *IEEE Software*, March 1989, pp. 46-51.

[58] Kleinrock, L., *Queuing Systems, Vol. 1: Theory*, Wiley Publishing, New York, 1975.

[59] Graham, G. S., "Queuing Network Models of Computer System Performance," *Computing Surveys*, Vol. 10, No. 3, September 1978, pp. 219-224.

[60] Allen, A. O., "Queuing Models of Computer Systems," *IEEE Computer*, April 1980, pp. 13-24.

[61] Balbo, G., S. C. Bruell, S. Ghanta, "Combining Queuing Networks and Generalized Stochastic Petri Nets for Solutions of Complex Models of System Behavior," *IEEE Transactions on Computer*, October 1988, pp. 1251-1268.

[62] Hady, F., J. H. Aylor, R. Waxman, B. W. Johnson, R. D. Williams, "Uninterpreted/Interpreted Modeling of Digital Systems in a Common Simulation Environment," Department of Electrical Engineering,

University of Virginia, Technical Report No. CSIS 881208.0, December 8, 1988.

[63] Smith, C. U., "Robust Models for the Performance Evaluation of Software/Hardware Design," *International Workshop on Timed Petri Nets*, Torino, Italy, July 1-3, 1985, pp. 172-180.

[64] Rao, R., *A Building Block Approach to Performance Modeling in VHDL*, Master's Thesis, University of Virginia, May 1990.

[65] Antoniazzi, S., M. Mastretti, "An Interactive Visual Environment for Hardware/Software System Design at the Specification Level," *Microprocessing and Microprogramming 30*, 1990, pp. 545-554.

[66] Schriber, J. J., *Simulation using GPSS*, John Wiley & Sons, New York, 1974.

[67] Kiviat, P. J., R. Villanueva, H. M. Markowitz, "*SIMSCRIPT II.5 Programming Language*," CACI, Los Angeles, Ca., 1973.

[68] Iacobovici, S., C. Ng, "VLSI and System Performance Modeling," *IEEE Micro*, August 1987, pp. 59-72.

[69] Gordon, R. F., E. A. McNair, P. D. Welch, "Examples of Using the Research Queuing Package Modeling Environment (RESQME)," *Proceedings 1986 Winter Simulation Conference*, 1986, pp. 494-503.

[70] Melamed, B., R. J. T. Morris, "Visual Simulation: The Performance Analysis Workstation," *IEEE Computer*, Vol. 18, August 1985, pp. 87-94.

[71] Funka-Lea, C. A., T. D. Kontogiorgos, R. J. T. Morris, L. D. Rubin, "Interactive Visual Modeling for Performance," *IEEE Software*, September 1991, pp. 58-69.

[72] Rose, C. W., M. Buchner, Y. Trivedi, "Integrating Stochastic Performance Analysis with System Design Tools," *Proceedings 22nd Design Automation Conference*, 1985, pp. 482-488.

[73] Rose, C. W., "The What and How of Top Down Design," TD

Technologies, April 1992.

[74] Vernon, M. K., G. Estrin, "The UCLA Graph Model of Behavior: Support for Performance-Oriented Design," *Proceedings IFIP WG10.1 Working Conference on Methodology for Computer System Design*, Lille, France, September 1983, pp. 47-65.

[75] Estrin, G., R. S. Fenchel, R. R. Razouk, M. K. Vernon, "SARA: Modeling, Analysis, and Simulation Support for Design of Concurrent Systems," *IEEE Transactions on Software Engineering*, Vol. SE-12, No. 2, February 1986, p. 293-311.

[76] Scientific Engineering Software, Inc., *SES/Workbench User's Guide*, Austin, Texas, April 1989.

[77] Sholl, H. A., T. L. Booth, "Software Performance Modeling using Computation Structures," *IEEE Transactions on Software Engineering*, Vol. SE-1, No. 4, December 1975, pp. 414-420.

[78] Smith, C., J. C. Browne, "Performance Specifications and Analysis of Software Designs," *Conference on Simulation, Measurement and Modeling of Computer Systems*, 1979, pp. 173-182.

[79] Booth, T. L., "Performance Optimization of Software Systems Processing Information Sequences Modeled by Probabilistic Languages," *IEEE Transactions on Software Engineering*, Vol. SE-5, No. 1, January 1979, pp. 31-44.

[80] Booth, T. L., C. A. Wiecek, "Performance Abstract Data Types as a Tool in Software Performance Analysis," *IEEE Transactions on Software Engineering*, Vol. SE-6, No. 2, March 1980, pp. 138-151.

[81] Oldehoeft, R. R., "Program Graphs and Execution Behavior," *IEEE Transactions on Software Engineering*, Vol. SE-9, No. 1, January 1983, pp. 103-108.

[82] Gomaa, H., "A Software Design Method for Real-Time Systems," *Communications of the ACM*, Vol. 27, No. 9, September 1984, pp. 938-949.

[83] Ammar, R. A., B. Qin, "An Approach to Derive Time Costs for Sequential Computations," *Journal Systems Software*, Vol. 11, 1990, pp. 173-180.

[84] Ammar, R. A., "A Computer Aided Design System to Develop High Performance Software," *Journal Systems Software*, Vol. 15, 1991, pp. 139-147.

[85] Cheung, R. C., "A User-Oriented Software Reliability Model," *IEEE Transactions on Software Engineering*, Vol. SE-6, No. 2, March 1980, pp. 118-125.

[86] Mok, D. S., S. T. Becker, "Simulating a Complex Software System," *Annual Simulation Symposium*, 1990, pp. 33-49.

[87] Ammar, R. A., M. M. Farid, K. Yetongnon, "A Spreadsheet Performance Approach to Integrate a Modeling Hierarchy of Software Systems," *International Conference Systems, Man, and Cybernetics*, 1989, pp. 847-852.

[88] Littlewood, B., "Theories of Software Reliability: How Good are they and How can they be Improved?," *IEEE Transactions on Software Engineering*, Vol. SE-6, No. 5, September 1980, pp. 489-500.

[89] Goel, A. L., "Software Reliability Models: Assumptions, Limitations, and Applicability," *IEEE Transactions on Software Engineering*, Vol. SE-11, No. 12, December 1985, pp. 1411-1423.

[90] Peterson, J. L., *Petri Net Theory and the Modeling of Systems*, Englewood Cliffs, N.J., Prentice-Hall, 1981.

[91] Kavi, K. M., B. P. Buckles, U. N. Bhat, "Isomorphisms between Petri nets and Dataflow Graphs," *IEEE Transactions on Software Engineering*, Vol. SE-13, No. 10, October 1987, pp. 1127-1133.

[92] Auletta, R. J., *An Uninterpreted Model for Hardware Description Languages*, Department of Electrical Engineering, Ph. D. Dissertation, University of Virginia, May 1987.

[93] Chu, W. W., L. J. Holloway, M. Lan, K. Efe, "Task Allocation in

Distributed Data Processing," *IEEE Computer*, November 1980, pp. 57-69.

[94] Kruatrachue, G., T. Lewis, "Grain Size Determination for Parallel Processing," *IEEE Software*, January 1988, pp. 23-32.

[95] Stone, H. S., "Multiprocessor Scheduling with the Aid of Network Flow Algorithms," *IEEE Transactions on Software Engineering*, Vol. SE-3, No. 1, January 1977, pp. 85-93.

[96] Bokhari, S. H., "On the Mapping Problem," *IEEE Transactions on Computers*, Vol. C-30, No. 3, March 1981, pp. 207-214.

[97] Pathak, G. C., "Towards Automated Design of Multicomputer System for Real-Time Applications," Ph.D. Dissertation, N. C. State University, July 1984.

[98] Jalote, P., *An Integrated Approach to Software Engineering*, Springer-Verlag, New York, 1991.

[99] Davis, A. M., "A Comparison of Techniques for the Specification of External System Behavior," *Communications of the ACM*, Vol. 31, No. 9, September 1988, pp. 1098-1115.

[100] Wang, Y., "A Distributed Specification Model and its Prototyping," *IEEE Transactions on Software Engineering*, Vol. 14, No. 8, August 1988, p. 1090-1097.

[101] Mills, H. D., "Stepwise Refinement and Verification in Box-Structured Systems," *IEEE Computer*, June 1988, pp. 23-36.

[102] Gabrielian, A., M. K. Franklin, "Multi-level Specification and Verification of Real-Time Software," *12th International Conference on Software Engineering*, 1990, pp. 52-62.

[103] Ross, D., "Structured Analysis (SA): A Language for Communicating Ideas," *IEEE Transactions on Software Engineering*, Vol. SE-3, No. 1, January 1977.

[104] DeMarco, T., *Structured Analysis and Specification*, Yourdon

Press, New York, 1978.

[105] Bruno, G., G. Marchetto, "Process-Translatable Petri Nets for the Rapid Prototyping of Process Control Systems," *IEEE Transactions on Software Engineering*, Vol. SE-12, No. 2, February 1986, pp 346-357.

[106] Oswald, H., R. Esser, R. Mattmann, "An Environment for Specifying and Executing Hierarchical Petri nets," *12th International Conference on Software Engineering*, 1990, pp. 164-171.

[107] Zave, P., "The Operational versus the Conventional Approach to Software Development," *Communications of the ACM*, Vol. 27, No. 2, February 1984, pp. 104-118.

[108] Zave, P., W. Schell, "Salient Features of an Executable Specification Language and its Environment," *IEEE Transactions on Software Engineering*, Vol. SE-12, No. 2, February 1986, pp 312-325.

[109] Harel, D., "STATEMATE: A Working Environment for the Development of Complex Reactive Systems," *IEEE Transactions on Software Engineering*, Vol. 16, No. 4, April 1990, pp. 403-414.

[110] Yourdon, E., L. Constantine, *Structured Design*, Prentice-Hall, Englewood Cliffs, New York, 1979.

[111] Hoare, C. A. R., "An Axiomatic Basis for Computer Programming," *Communications of the ACM*, Vol. 12, No. 3, March 1969, pp. 335-355.

[112] Linger, R. C., H. D. Mills, B. I. Witt, *Structured Programming Theory and Practice*, Addison-Wesley Publishing, Reading, Massachusetts, 1979.

[113] Boehm, B. W., *Software Engineering Economics*, Prentice-Hall, Englewood Cliffs, N.J., 1981.

[114] Basili, V. R., A. Turner, "Iterative Enhancement, a Practical Technique for Software Development," *IEEE Transactions on Software Engineering*, Vol. SE-1, No. 4, December 1975.

[115] Boehm, B. W., "A Spiral Model of Software Development and Enhancement," *IEEE Computer*, May1988, pp. 61-72.

[116] Boehm, B. W., "Software Risk Management: Principles and Practices," *IEEE Software*, January 1991, pp. 32-41.

[117] Liskov, B., J. Guttag, *Abstraction and Specification in Program Development*, MIT Press, Cambridge, Massachusetts, 1986.

[118] Dijkstra, E. W., "Structured Programming," in *Software Engineering Techniques*, NATO Science Committee, J. N. Buxton and B. Randell, eds., 1969, pp. 84-88.

[119] Jensen, R. W., "Structured Programming," *IEEE Computer*, March 1981, pp. 31-48.

[120] Tausworthe, R. C., "Structured Programming and Software Engineering of Hard Real-Time Minicomputer Systems," *11th Annual Asilomar Conference on Circuits, Systems, and Computers*, 1978, pp. 289-294.

[121] Liskov, B., "A Design Methodology for Reliable Software Systems," *AFIPS Conference Proceedings*, Part I, 1972, pp. 191-199.

[122] Davis, A. L., R. M. Keller, "Data Flow Program Graphs," *IEEE Computer*, 1982, pp. 26-41.

[123] Stevens, W. P., "Using Data Flow for Application Development," *BYTE*, June 1985, pp. 267-276.

[124] Browne, J. C., M. Azam, S. Sobek, "CODE: A Unified Approach to Parallel Programming," *IEEE Software*, 1989, pp. 10-18.

[125] Stevens, W. P., G. J. Myers, L. L. Constantine, "Structured Design," *IBM Systems Journal*, Vol. 14, No. 2, 1974, pp. 115-139.

[126] Wirth, N., "Program Development by Stepwise Refinement," *Communications of the ACM*, Vol. 14, No. 4, April 1971, pp. 221-227.

[127] McClure, C. L., "Top-Down, Bottom-Up, and Structured Programming," *IEEE Transactions on Software Engineering*, Vol. SE-

1, No. 4, December 1975.

[128] Sommerville, I., *Software Engineering*, 3rd Edition, Addison Wesley, 1989.

[129] Wulf, W. A., R. L. London, M. Shaw, "An Introduction to the Construction and Verification of Alphard Programs," *IEEE Transactions on Software Engineering*, Vol. SE-2, No. 4, December 1976.

[130] Guttag, J., "Abstract Data Types and the Development of Data Structures," *Communications of the ACM*, Vol. 20, No. 6, June 1977.

[131] Parnas, D. L., "On the Criteria to be used in Decomposing Systems into Modules," *Communications of the ACM*, Vol. 15, No. 12, December 1972, pp. 1053-1058.

[132] Parnas, D. L., "The Modular Structure of Complex Systems," *7th International Conference on Software Engineering*, 1984, pp. 408-417.

[133] Robson, D., "Object-Oriented Systems," *BYTE*, August 1981, pp. 74-86.

[134] Cox, B., "Message/Object Programming: An Evolutionary Change in Programming Technology," *IEEE Software*, January 1984, pp. 50-61.

[135] Booch, G., "Object-Oriented Development," *IEEE Transactions on Software Engineering*, Vol. SE-12, No. 2, February 1986, pp. 211-221.

[136] Pascoe, G., "Elements of Object-Oriented Programming," *BYTE*, August 1986, pp. 139-144.

[137] Meyer, B., "Reusability: The Case for Object-Oriented Design," *IEEE Software*, March 1987, pp. 50-64.

[138] Ward, P. T., "How to Integrate Object Orientation with Structured Design and Analysis," *IEEE Software*, March 1989, pp. 74-82.

[139] Wasserman, A. I., P. A. Pircher, R. J. Muller, "The Object-

Oriented Structured Design Notation for Software Design Representation," *IEEE Computer*, March 1990, pp. 50-63.

[140] Setliff, D., E. Kant, T.Cain, "Practical Software Synthesis," *IEEE Software*, May 1993, pp. 6-10.

[141] Smith, D. R., "KIDS: A Semiautomatic Program Development System," *IEEE Transactions on Software Engineering*, Vol. 16, No. 9, September 1990, pp. 1024-1043.

[142] Rich, C., R. C. Waters, "Automatic Programming: Myths and Prospects," *IEEE Computer*, August 1988, pp. 40-51.

[143] Jüllig, R. K., "Applying Formal Software Synthesis," *IEEE Software*, May 1993, pp. 11-22.

[144] Kant, E., "Synthesis of Mathematical-Modeling Software," *IEEE Software*, May 1993, pp. 30-41.

[145] Abbott, B., T. Bapty, C. Biegl, G. Karsai, J. Sztipanovits, "Model-Based Software Synthesis," *IEEE Software*, May 1993, pp. 42-52.

[146] Myers, G. J., *Advances in Computer Architecture*, John Wiley and Sons, New York, 1982.

[147] Rose, J., A. El Gamal, A. Sangiovanni-Vincentelli, "Architecture of Field-Programmable Gate Arrays," *Proceedings of the IEEE*, Vol. 81, No. 7, July 1993, pp. 1013-1029.

[148] Camposano, R., "From Behavior to Structure: High-Level Synthesis," *IEEE Design & Test*, October 1990.

[149] Dutt, N., D. D. Gajski, "Design Synthesis and Silicon Compilation," *IEEE Design & Test*, October 1990.

[150] Sarma, R. C., M. D. Dooley, N. C. Newman, G. Hetherington, "High-level Synthesis: Technology Transfer to Industry," *Proceedings 27th Design Automation Conference*, 1990.

[151] McFarland, M. C., "Using Bottom-Up Design Techniques in the Synthesis of Digital Hardware from Abstract Behavioral Descriptions,"

Proceedings 23th Design Automation Conference, 1986.

[152] Thomas, D. E., E. D. Lagnese, R. A. Walker, J. A. Nestor, J. V. Rajan, R. L. Blackburn, *Algorithmic and Register-Transfer Level Synthesis: The System Architect's Workbench,* Kluwer Academic Publishers, 1990.

[153] De Micheli, G., D. C. Ku, "HERCULES: A System for High Level Synthesis," *Proceedings of the 25th Design Automation Conference*, June 1988, pp. 483-488.

[154] Gregory, D. K. Bartlett, A. De Geus, "SOCRATES: A System for Automatically Synthesizing and Optimizing Combinational Logic," *Proceedings 23th Design Automation Conference*, 1986, pp 79-85.

[155] De Micheli, G., "The High-Level Synthesis of Digital Circuits," *IEEE Design & Test*, October 1990.

[156] Kowalski, T., "The VLSI Design Automation Assistant: From Algorithms to Silicon," *IEEE Design & Test*, August 1985.

[157] Paulin, P. G., J. P. Knight, E. F. Girczyc, "HAL: A Multi-Paradigm Approach to Automatic Data Path Synthesis," *Proceedings 23th Design Automation Conference*, 1986, pp 263-270.

[158] Park, N., A. C. Parker, "SEHWA: A Program for Synthesis of Pipelines," *Proceedings 23th Design Automation Conference*, 1986, pp 454-460.

[159] Marwedel, P., "A New Synthesis Algorithm for the MIMOLA Software System," *Proceedings 23th Design Automation Conference*, 1986, pp 271-277.

[160] Parker, A. C., J. Pizarro, M. Milnar, "MAHA: A Program for Data Path Synthesis," *Proceedings 23th Design Automation Conference*, 1986, pp 461-466.

[161] Breternitz Jr., M., J. P. Shen, "Architecture Synthesis of High-Performance Application-Specific Processors," *Proceedings 27th Design Automation Conference*, 1990, pp 542-548.

[162] De Man, H., J. Rabaey, P. Six, L. Claesen, "Cathedral II: A Silicon Compiler for Digital Signal Processing," *IEEE Design & Test*, December 1986.

[163] Haroun, B. S., M. I. Elmasry, "Architectural Synthesis for DSP Silicon Compilers," *IEEE Transactions on Computer-Aided Design*, April 1989.

[164] Jain, P. P., S. Dhingra, J. C. Browne, "Bringing Top-Down Synthesis into the Real World," *High Performance Systems*, Vol. 10, July 1989, pp. 86-94.

[165] McFarland, M. C., "Reevaluating the Design Space of Register-Transfer Hardware Synthesis," *Proceedings of ICCAD*, November 1987, pp. 262-265.

[166] De Micheli, G., D. Ku, F. Mailhot, T. Truong, "The Olympus Synthesis System," *IEEE Design & Test*, October 1990.

[167] Gupta, R. K., G. De Micheli, "Partitioning of Functional Models of Synchronous Digital Systems," *Proceedings of the International Conference on Computer-Aided Design,* 1990, p. 152-155.

[168] Lagnese, E. D., D. E. Thomas, "Architectural Partitioning for System Level Synthesis of Integrated Circuits," *IEEE Transactions of Computer-Aided Design*, Vol. 10, No. 7, July 1991.

[169] Vahid, F., D. D. Gajski, "Specification Partitioning for System Design," *29th Design Automation Conference*, June 1992, p. 219-224.

[170] Stallings, W., *Computer Organization and Architecture: Principles of Structure and Function*, Macmillan Publishing Co., NY, 1990.

[171] Flynn, M. J., R. I. Winner, "ASIC Microprocessors," *Proceedings of 22nd Annual Workshop on Microprogramming and Microarchitecture*, Dublin, Ireland, August 1989, pp. 237-243.

[172] Parker, A. C., A. W. Nagle, "Hardware/Software Tradeoffs in a Variable Word Width, Variable Queue Length Buffer Memory," *4th*

Annual Symposium on Computer Architecture, 1977, pp. 159-163.

[173] Chandy, K. M., C. V. Ramamoorthy, A. Cowan, "A Framework for Hardware-Software Tradeoffs in the Design of Fault-Tolerant Computers," *AFIPS Fall Joint Conference*, Part I, 1972, p. 55-63.

[174] Conklin, P. F., D. P. Rodgers, "Advanced Minicomputer Designed by Team Evaluation of Hardware/Software Tradeoffs," *Computer Design*, April 1978, pp. 129-137.

[175] Hennessy, J., N. Jouppi, F. Baskett, T. Gross, J. Gill, "Hardware/ Software Tradeoffs for Increased Performance," *Architectural Support for Programming Languages and Operating Systems*, 1982, p. 2-11.

[176] Malinowski, C. W., P. S. Danile, "Hardware-Software Trade-offs in Real-Time Systems," *VLSI Systems Design*, June 1988, p. 80-89.

[177] Rao, G. S., P. L. Rosenfeld, "Integration of Machine Organization and Control Program Design - Review and Direction," *IBM Journal of Research and Development*, Vol. 27, No. 3, May 1983, p. 247-256.

[178] Stockenberg, J. A. van Dam, "Vertical Migration for Performance Enhancement in Layered Hardware/Firmware./Software Systems," *IEEE Computer*, May 1978, pp. 33-50.

[179] Kamibayashi, N., H. Ogawana, K. Nagayama, H. Aiso, "HEART: An Operating System Nucleus Machine Implemented in Firmware," *Architectural Support for Programming Languages and Operating Systems*, 1982, pp. 195-204.

[180] Philipson, L., "Multilevel Design and Verification of Hardware/ Software Systems," *IEEE Journal of Solid-State Circuits*, Vol. 25, No. 3, June 1990, p. 714-719.

[181] Smith, C. U., R. R. Gross, "Technology Transfer between VLSI Design and Software Engineering: CAD Tools and Design Methodologies," *Proceedings of the IEEE*, Vol. 74, No. 6, June 1986, pp. 875-885.

[182] Finkelstein, A., B. Nuseibeh, "Technology Transfer: Software

Engineering and Engineering Design," *IEE Computing and Control Engineering Journal*, 3(6), November 1992, pp. 259-265.

[183] Brooks, F. P., "No Silver Bullet: Essence and Accidents of Software Engineering," *IEEE Computer*, Vol. 20, No. 4, April 1987, pp. 10-19.

[184] Wulf, W. A., *A Notation for Digital Systems*, Ph.D. Dissertation, University of Virginia, 1968.

[185] Biewald, J., P. Goehner, R. Lauber, H. Schelling, "EPOS - A Specification and Design Technique for Computer Controlled Real-Time Automation System," *Proceedings 4th International Conference on Software Engineering*, 1977, p. 245-250.

[186] Buchenrieder, K., "Codesign and Concurrent Engineering," in Hot Topics of *IEEE Computer*, R. D. Williams, ed., pp. 84-87.

[187] Ganapathi, M., C. N. Fischer, "Attributed Linear Representations for Retargetable Code Generation," *Software-Practice and Experience*, Vol. 14, April 1984, pp. 347-364.

[188] Chiodi, M., A. L. Sangiovanni-Vincentelli, "Design Methods for Reactive Real-Time Systems Co-Design," International Workshop on Hardware/Software Codesign, Estes Park, Colorado, September 1992.

[189] Bertrand, M., *Object-oriented Software Construction*, Prentice-Hall, New York, 1988.

[190] Baker, W., A. R. Newton, "Synchronous Parallelism and Object-oriented Computing for Real-Time Software Applications," International Workshop on Hardware/Software Codesign, Estes Park, Colorado, September 1992.

[191] Benders, L. P. M., M. P. J. Stevens, "Task Level Behavioral Description Translation to IEEE VHDL," *VHDL Forum for CAD in Europe*, Marseille, France, April 1991.

[192] Hoare, C. A. R., "Communicating Sequential Processes," *Communications of the ACM*, Vol. 21, No. 8, August 1978, pp. 666-677.

[193] Christ-Neumann, M. L., R. Budde, H. Nieters, M. Pinna, A. Pawlak, A. Poigné, K. Sylla, R. Camposano, "On an Experiment in System Codesign: A Mass Flowmeter," International Workshop on Hardware/Software Codesign, Estes Park, Colorado, September 1992.

[194] Franke, D. W., M. K. Purvis, "Hardware/Software Codesign Project Overview," Microelectronics and Computer Technology Corporation.

[195] Franke, D. W., M. K. Purvis, "An Overview of Hardware/ Software Codesign," *International Symposium on Circuits & Systems*, May 1992, pp. 2665-2668.

[196] Woo, N., W. Wolf, A. Dunlop, "Compilation of a Single Specification into Hardware and Software," International Workshop on Hardware/Software Codesign, Estes Park, Colorado, September 1992.

[197] Feather, M. S., S. Fickas, B. R. Helm, "Specification and Design of Composite Systems," International Workshop on Hardware/Software Codesign, Estes Park, Colorado, September 1992.

[198] Eles, P., Z. Peng, A. Doboli, "VHDL System-Level Specification and Partitioning in a Hardware/Software Co-Synthesis Environment," *3rd International Workshop on Hardware/Software Codesign*, Grenoble, France, September 22-24, 1994, pp. 49-55.

[199] Ernst, R., J. Henkel, T. Benner, "Hardware-Software Cosynthesis for Microcontrollers," *IEEE Design and Test*, December 1993, pp. 64-75.

[200] Gupta, R. K., G. De Micheli, "System-level Synthesis using Re-programmable Components," *Proceedings of the European Design Automation Conference*, March 1992, pp. 2-7.

[201] Ernst, R., J. Henkel, "Hardware-Software Codesign of Embedded Controllers Based on Hardware Extraction," International Workshop on Hardware/Software Codesign, Estes Park, Colorado, September 1992.

[202] Athanas, P., H. F. Silverman, "Processor Reconfiguration through Instruction-Set Metamorphosis," *IEEE Computer*, Vol. 26, No. 3,

March 1993, pp. 11-18.

[203] Barros, E., W. Rosenstiel, "A Method for Hardware/Software Partitioning," *Proceedings Compeuro*, IEEE CS Press, 1992.

[204] Kalavade, A., E. Lee, "A Global Criticality/Local Phase Driven Algorithm for the Constrained Hardware/Software Partitioning Problem," *3rd International Workshop on Hardware/Software Codesign*, Grenoble, France, September 22-24, 1994, pp. 42-48.

[205] Lee, R., "Empirical Results on the Speed, Efficiency, Redundancy, and Quality of Parallel Computations," International Conference on Parallel Processing, August 1980, pp. 91-100.

[206] Siegel, L. J., H. J. Siegel, P. H. Swain, "Performance Measures for Evaluating Algorithms for SIMD Machines," *IEEE Transactions on Software Engineering*, Vol. SE-8, No. 4, July 1982, pp. 319-330.

[207] Gupta, R. K., G. De Micheli, "Hardware-Software Cosynthesis for Digital Systems," *IEEE Design and Test*, September 1993, pp. 29-40.

[208] Smith, C. U., G. A. Frank, J. L. Cuadrado, "An Architecture Design and Assessment System for Software/Hardware Codesign," *Proceedings 22nd Design Automation Conference*, 1985, pp. 417-424.

[209] Acosta, R. D., "Use of Dataflow Specifications for Software/Hardware Codesign," International Workshop on Hardware/Software Codesign, Estes Park, Colorado, September 1992.

[210] Srivastava, M. B., R. W. Broderson., "Rapid-Prototyping of Hardware and Software in a Unified Framework," *Proceedings of the International Conference on Computer-Aided Design,* 1991, pp. 152-155.

[211] Kalavade, A., E. A. Lee, "Hardware/Software Co-Design Using Ptolemy - A Case Study," International Workshop on Hardware/Software Codesign, Estes Park, Colorado, September 1992.

[212] Kalavade, A., E. A. Lee, "A Hardware-Software Codesign

Methodology for DSP Applications," *IEEE Design and Test*, September 1993, pp. 16-28.

[213] Chou, P., R. Ortega, G. Borriello, "Synthesis of the Hardware/ Software Interface in Microcontroller-Based Systems," International Workshop on Hardware/Software Codesign, Estes Park, Colorado, September 1992.

[214] Wenban, A. S., J. W. O' Leary, G. M. Brown, "Codesign of Communication Protocols," International Workshop on Hardware/ Software Codesign, Estes Park, Colorado, September 1992.

[215] Adams, J. K., D. E. Thomas, "Addressing the Tradeoff between Standard and Custom ICs in System Level Design," Hardware-Software Systems", International Workshop on Hardware/Software Codesign, Estes Park, Colorado, September 1992.

[216] Gupta, R. K., C. N. Coelho Jr., G. De Micheli, "Synthesis and Simulation of Digital Systems Containing Interacting Hardware and Software Components," *29th Design Automation Conference*, June 1992, pp. 225-230.

[217] Gupta, R. K., C. N. Coelho, Jr., G. De Micheli, "Program Implementation Schemes for Hardware-Software Systems," International Workshop on Hardware/Software Codesign, Estes Park, Colorado, September 1992.

[218] Becker, D., R. K. Singh, S. G. Tell, "An Engineering Environment for Hardware/Software Co-Simulation," *29th Design Automation Conference*, June 8-12, 1992, pp. 129-134.

[219] Adams, J. K., H. Schmitt, D. E. Thomas, "A Model and Methodology for Hardware-Software Codesign," International Workshop on Hardware-Software Co-design, Cambridge, Massachusetts, October 7-8, 1993.

[220] Hersen, R., "Charon - A Co-Simulation Application," Second IFIP International Workshop on Hardware/Software Codesign, Innsbruck, Austria, May 24-27, 1993.

[221] Ostman, F., P. Gibson, "Early Integration in Industrial Practise," Second IFIP International Workshop on Hardware/Software Codesign, Innsbruck, Austria, May 24-27, 1993.

[222] Van Dun, J., "HdS/H Cosim: A Cosimulation Prototype Applied in the Formal Design of Telecom PBAs," Second IFIP International Workshop on Hardware/Software Codesign, Innsbruck, Austria, May 24-27, 1993.

[223] Hagen, K., H. Meyr, "Timed and Untimed Hardware/Software Co-Simulation: Application and efficient Implementation," International Workshop on Hardware-Software Co-design, Cambridge, Massachusetts, October 7-8, 1993.

[224] Thomas, D. E., J. K. Adams, H. Schmitt, "A Model and Methodology for Hardware-Software Codesign," *IEEE Design and Test*, September 1993, pp. 6-15.

[225] Billowitch, W., "Simulation Models support HW/SW Integration," *Computer Design*, March 1, 1988.

[226] Kern, A., A. Blazevicius, "A Concurrent Hardware and Software Design Environment," *VLSI Systems Design*, August 1988, pp. 34-40.

[227] Buchenrieder, K., C. Veith, "CODES: A Practical Concurrent Design Environment," International Workshop on Hardware/Software Codesign, Estes Park, Colorado, September 1992

[228] Liu, C. L., J. W. Layland, "Scheduling Algorithm for Multiprogramming in a Hard Real-Time Environment," *Journal of the ACM*, Vol. 20, No. 1, January 1973.

[229] Cochran, M., "Using the Rate Monotonic Analysis to Analyze the Schedulability of ADARTS Real-Time Software Designs," International Workshop on Hardware/Software Codesign, Estes Park, Colorado, September 1992.

[230] Shen, V. Y., C. Richter, M. L. Graf, J. A. Brumfield, "VERDI: A Visual Environment for Designing Distributed Systems," *Journal of Parallel and Distributed Computing*, 1990, p. 128-137.

[231] Huang, I., A. Despain, "High-Level Synthesis of Pipelined Instruction Set Processors and Back-End Compilers," *Proceedings 29th Design Automation Conference*, June 8-12 1992, pp. 135-140.

[232] Plankl, J., K. Westerholz, "Balanced Systems: Integration of Hardware and Software Design," International Workshop on Hardware/ Software Codesign, Estes Park, Colorado, September 1992.

[233] Cook, T. A., E. A. Harcourt, T. K. Miller III, P. D. Franzon, "Towards Unified Specification for Architecture and Compiler Design," International Workshop on Hardware/Software Codesign, Estes Park, Colorado, September 1992.

[234] Jain, P. P, S. Dhingra, J. C. Browne, "Bringing Top Down Synthesis into the Real World," *High Performance Systems*, Vol. 10, July 1989, pp. 86-94.

[235] JRS Research Laboratories, Inc., Customer documentation.

[236] Srinivasan, S., *ADEPT: An Advanced Design Environment Prototyping Tool*, Master's Thesis, University of Virginia, May 1990.

[237] Kumar, S., R. H. Klenke, J. H. Aylor, B. W. Johnson, R. D. Williams, R. Waxman, "ADEPT: A Unified System Level Modeling Design Environment," *Proceedings of the 1st Annual RASSP Conference* (non-ITAR restricted), August 15-18, 1994, pp. 114-123.

[238] Jensen, K., "Colored Petri Nets: A High Level Language for System Design and Analysis," in *High-level Petri Nets: Theory and application,* K. Jensen and G. Rozenberg (Eds.), Berlin: Springer-Verlag, 1991, pp. 44-119.

[239] Kumar, S., J. H. Aylor, B. W. Johnson, W. A. Wulf, "A Framework for Hardware/Software Codesign," *IEEE Computer*, Vol. 26, No. 12, December 1993, pp.39-45.

[240] Prather, R. E., *Discrete Mathematical Structures for Computer Science*, Houghton Mifflin Publishing Company, Boston, 1976.

[241] Kumar, S., J. H. Aylor, B. W. Johnson, W. A. Wulf, "A

Framework for Hardware/Software Codesign," International Workshop on Hardware/Software Codesign, Estes Park, Colorado, September 1992.

[242] Private Conversation, Dr. Ronald D. Williams, Professor, Electrical Engineering, University of Virginia.

[243] Private Conversation, Dr. Ronald D. Williams, Professor, Electrical Engineering, University of Virginia.

[244] Kumar, S., J. H. Aylor, B. W. Johnson, W. A. Wulf, "A Framework for Hardware/Software Codesign," University of Virginia, Technical Report, May 1992.

[245] Kumar, S., J. H. Aylor, B. W. Johnson, W. A. Wulf, R. D. Williams, "An Abstract Hardware/Software Model for Early Performance Evaluation," *Proceedings of the Symposium and Workshop on Systems Engineering of Computer Based Systems (SECBS '95)*, Tucson, Arizona, March 1995, pp. 299-306 (Errata: Reference [6] in this paper should be the same as Reference [8] in the paper).

[246] Gilchrist, W., *Statistical Modelling*, John Wiley and Sons, New York, 1984.

[247] Wulf, W., M. Shaw, P. N. Hilfinger, L. Flon, *Fundamental Structures of Computer Science*, Addison-Wesley Publishing Company, Reading, Massachusetts, 1981.

[248] MacDonald, R., R. D. Williams, J. H. Aylor, "An Approach to Unified Performance and Functional Modeling of Complex Systems," *IASTED International Conference on Modeling and Simulation*, Pittsburgh, Pennsylvania, April 1995.

[249] Agerwala, T. K. M., *Towards a Theory for the Analysis and Synthesis of Systems Exhibiting Concurrency*, Ph. D. Dissertation, Dept. of Electrical Engineering and Computer Science, The Johns Hopkins University, 1975.

[250] Kumar, S., J. H. Aylor, B. W. Johnson, W. A. Wulf, "Object-Oriented Techniques in Hardware Design," *IEEE Computer*, Vol. 27,

No. 6, June 1994, pp.64-70.

[251] Huck, J. C., M. J. Flynn, *Analyzing Computer Architectures*, IEEE Computer Society Press, Washington, D. C., 1989.

[252] Flynn, M. J., "Directions and Issues in Architecture and Language," *IEEE Computer*, October 1980, pp. 5-22.

[253] Hoffman, R., "A Classification of Interpreter Systems," *Microprocessing and Microprogramming*, 12, 1983, pp. 3-8.

[254] Ammann, U., "On Code Generation in a Pascal Compiler," *Software-Practice and Experience*, June/July 1977, pp. 391-423.

[255] Perkins, D. R., R. L. Sites, "Machine-Independent Pascal Code Optimization," *Proceedings ACM SIGPLAN Symposium Compiler Construction*, Denver, Colorado, August 6-10, 1979, pp. 201-207.

[256] Debaere, E. H., J. M. Van Campenhout, *Interpretation and Instruction Path Coprocessing*, MIT Press, Cambridge, Massachusetts, 1990.

[257] Stankovic, J. A., "The Types and Interactions of Vertical Migrations of Functions in a Multilevel Interpretive System," *IEEE Transactions on Computers*, Vol. C-30, No. 7, July 1981.

[258] Flynn, M. J., "Some Computer Organizations and Their Effectiveness," *IEEE Transactions on Computers*, Vol. C-21, No. 9, September 1972, pp. 948-960.

[259] Frank, G., J. DiSanto, "Software/Hardware Codesign of Real-Time Systems with ADAS," *Electronic Engineering*, March 1990, pp. 95-102.

[260] Aho, A. V., R. Sethi, J. D. Ullman, *Compilers Principles, Techniques, and Tools*, Addison-Wesley Publishing Company, Reading, Massachusetts, 1986.

[261] Böhm, C., G. Jacopini, "Flow Diagrams, Turing Machines, and Languages with Only Two Formulation Rules," *Communications of the*

ACM, Vol. 9, No. 5, May 1966, pp. 366-371.

[262] Miller, R. E., "A Comparison of Some Theoretical Models of Parallel Computation," *IEEE Transactions on Computer*, Vol. C-22, No. 8, August 1973, pp. 710-717.

[263] Developed through private discussions with Dr. Ronald D. Williams, Professor, Electrical Engineering, University of Virginia.

[264] Shaikh, A., "Design of Input and Output Modules for a Safety-Critical Wayside Train Control System," Master's Thesis, University of Virginia, December, 1994.

[265] Perrone, P. J., "Global Safety Assurance: Concepts and Application to Train Control Systems," Master's Thesis, University of Virginia, May 1995.

[266] Miller, W. H., B. W. Johnson, "Automatic Classification of Aluminum Defects," Department of Electrical Engineering, University of Virginia, Technical Report No. UVA/5-38459/EE94, October 31, 1994.

[267] Schaefer, P., R. D. Williams, "The Stylus Tracking Project," Department of Electrical Engineering, University of Virginia, Technical Report, 1995.

[268] Owen, R. E., "A 15 Nanosecond Complex Multiplier-Accumulator for FFTs," *ICASSP '87*, 1987, pp. 527-530.

[269] Fujimoto, R. M., "Parallel Discrete Event Simulation," *Communications of the ACM*, Vol. 33, No. 10, October 1990, pp. 30-53.

[270] Reynolds, Jr., P. F., C. M. Pancerella, S. Srinivasan, "Design and Performance Analysis of Hardware Support for Parallel Simulations," *Journal of Parallel and Distributed Computing,* Vol. 18, August 1993, pp. 435-453.

[271] Reynolds, Jr., P. F., "An Efficient Framework for Parallel Simulations," *International Journal in Computer Simulation 2,* 1992, pp. 427-445.

[272] Stroustrup, B., *The C++ Programming Language,* 2nd Edition, Addison-Wesley Publishing Company, Reading, Massachusetts, 1991.

[273] Tobias, J. R., "LSI/VLSI Building Blocks," *IEEE Computer*, August 1981, pp. 83-101.

[274] Peels, A. J. H. M., "Designing Digital Systems - SSI and MSI vs. LSI and VLSI," *IEEE Micro*, April 1987, pp. 66-80.

[275] Gnu C++ library.

[276] Wynia, T., "RISC and CISC Processors Target Embedded Systems," *Electronic Design*, June 27, 1991, pp. 55-70.

[277] Wolfe, A., J. P. Shen, "Flexible Processors: A Promising Application-Specific Processor Design Approach," *Proceedings of the 21st Annual Workshop on Microprogramming and Microarchitectures*, 1988, pp. 30-39.

[278] Mulder, H., P. Stravers, "A Flexible VLSI Core for an Adaptable Architecture," *Proceedings of the 22nd Annual Workshop on Microprogramming and Microarchitectures*, 1989, pp. 223-231.

[279] Booch, G., *Object-Oriented Analysis and Design with Applications*, 2nd Edition, Benjamin/Cummings Publishing Co., Inc., Redwood City, California, 1994.

[280] Williams, R. D. "Class Project Fall 1990," University of Virginia.

[281] Shaw, A. C., "Reasoning About Time in Higher-Level Language Software," *IEEE Transactions on Software Engineering*, Vol. 15, No. 7, July 1989, pp. 875-889.

[282] Davidson, J. W., J. V. Gresh, "Cint: A RISC Interpreter for the C Programming Language," *SIGPLAN '87 Symposium on Interpreters and Interpretive Techniques*, St. Paul, Minnesota, June 24-26, 1987, pp. 189-198.

[283] Meyer, B., "Genericity versus Inheritance," *Proceedings of the Object-Oriented Programming Systems, Languages, and Applications*

Conference (OOPSLA '86), September 1986, pp. 391-405.

[284] Cardelli, L., P. Wegner, "On Understanding Types, Data Abstraction, and Polymorphism," *Computing Surveys*, Vol. 17, No. 4, December 1985, pp. 471-522.

[285] Khoshafian, S., R. Abnous, *Object Orientation: Concepts, Languages, Databases, and User Interfaces*, John Wiley and Sons, Inc., New York, 1990.

[286] Parnas, D. L., "On the Design and Development of Program Families," *IEEE Transactions on Software Engineering*, Vol. SE-2, No. 1, March 1976, pp. 1-9.

[287] Giloi, W. K., "Towards a Taxonomy of Computer Architectures Based on the Machine Data Type View," *10th International Symposium on Computer Architecture*, 1983, pp. 6-13.

[288] Langdon, Jr., G. G., *Computer Design*, Computeach Press Inc., San Jose, Ca., 1982.

[289] Müller, W., F. Rammig, "ODICE: Object-Oriented Hardware Description in CAD Environment," *Proceedings of the Ninth International Symposium on Computer Hardware Description Languages and their Applications*, J. A. Darringer and F. J. Rammig, eds., Elsevier Science Publishers B. V. (North-Holland), IFIP, 1990, pp. 19-34.

[290] Chung, M. J., S. Kim, "An Object-Oriented VHDL Design Environment," *27th ACM/IEEE Design Automation Conference*, 1990, pp. 431-436.

[291] Nelson, M. L., K. A. Fontes, A. Zaky, "An Object-Oriented Approach to Computer Architecture Simulation," *Proceedings of the 25th Annual Hawaii International Conference on System Sciences (HICSS-25)*, Vol. 1: Architecture and Emerging Technologies, January 7-10, 1992, Kauai, Hawaii, pp. 476-485.

[292] Covnot, B. M., D. W. Hurst, S. Swamy, "OO-VHDL: An Object-Oriented VHDL," *1994 VHDL International User's Forum*.

[293] Newton, D. E., P. W. Vaughn, R. P. Johns, "PRISM: An Object-Oriented System Modeling Environment with an Embedded Symbolic Spreadsheet," *Proceedings of the 1991 (23rd) Summer Computer Simulation Conference*, July 22-24, 1991, Baltimore, Maryland.

[294] Yokote, Y., M. Tokoro, "Concurrent Programming in Concurrent Smalltalk," in *Object-Oriented Concurrent Programming*, A. Yonezawa and M. Tokoro, eds., MIT Press, Cambridge, Massachusetts, 1987.

[295] Wolf, W., "Object-Oriented Programming for CAD," *IEEE Design and Test of Computers*, March 1991, pp. 35-42.

[296] Gupta, R., W. H. Cheng, R. Gupta, I. Hardonag, M. A. Breuer, "An Object-Oriented VLSI CAD Framework," *IEEE Computer*, Volume 22, May 1989, pp. 28-37.

[297] Ledbetter, L., B. Cox, "Software-ICs," *BYTE*, June 1985, pp. 307-316.

[298] Jalote, P., "Synthesizing Implementations of Abstract Data Types from Axiomatic Specifications," *Software-Practice and Experience*, Vol. 17, No. 11, November 1987, pp. 847-858.

[299] Subrahmanyam, P. A., "Extending CAD Tools and Techniques," in Hot Topics of *IEEE Computer*, R. D. Williams, ed., January 1993, pp. 84-87.

[300] Welke, S., "A Unified Model of Hardware/Software Reliability," Master's Thesis, University of Virginia, May 1988.

[301] Grimshaw, A. S., "Easy-to-Use Object-Oriented Parallel Processing with Mentat," *IEEE Computer*, Vol. 26, No. 5, May 1993, pp. 39-51.

[302] Browne, J. C., J. Werth, T. Lee, "Intersection of Parallel Structuring and Reuse of Software Components: A Calculus of Composition of Components for Parallel Programs," *International Conference on Parallel Processing*, 1989, pp. 126-130.

Index